A HISTORY OF SEEING

A HISTORY OF SEEING

IN ELEVEN INVENTIONS

SUSAN DENHAM WADE

For Rob, Charlie, Stella,
Rosie and Hattie Boo

First published by The History Press as *As Far As The Eye Can See: A History of Seeing*, 2019

This updated paperback edition first published 2021

FLINT is an imprint of The History Press
97 St George's Place, Cheltenham,
Gloucestershire, GL50 3QB
www.flintbooks.co.uk

British Library Cataloguing in Publication Data.
A catalogue record for this book is available from the British Library.

ISBN 978 0 7509 9716 4

Typesetting and origination by The History Press
Printed and bound in Turkey by Imak.

CONTENTS

FOREWORD
BY TRISTAN GOOLEY

Without a brain, we would be little more than sacks of water, proteins and fats. And without our senses, our brain is tofu. Not literally, but as good as.

Our senses give our brain the information it needs to understand our world and make decisions about ways to improve our lives and avoid danger. The senses are the keys to a richer, more dynamic and safer life. Sight is the most powerful sense for more than 99 per cent of the population.

But there is a problem. We don't know what we're doing.

We muddle through life with a pair of super-tools bulging out of the front of our skull, hoping to learn how to use them as we go. We pick up some vague clues along the way by studying the behaviour of others. We learn that watching YouTube videos doesn't make us wise – it makes us fat.

Our eyes are the most extraordinary tools we will ever use, but they come with no manual. And most of us wouldn't read a book of dry instructions even if we were handed one. Fortunately, Denham Wade shows us what we need to know through the colourful lens of a cultural history of our relationship with this sense. And she brings this history into our lives vividly, bridging vast gaps so that we can see the past. We learn that individualism flourished soon after the first polished mirrors appeared in Turkey.

It is this weaving of world history and very personal history that thrills. Did you know that overweight people overestimate distances or that we see ourselves and our partners as better looking than we are? I didn't, but it did make me think. It doesn't apply to me, I'm sure, but it is very clever writing that tickles our weaknesses. We absolutely must find out how others see us: we are powerless against our own vanity.

In his book *Sapiens*, Yuval Noah Harari gave us a portrait of our broad family history. *A History of Seeing in Eleven Inventions* paints a picture that is more intimate, closer both physically and in time.

After reading this book, I could see how things were not as they first looked. You'll view things differently too.

PREFACE

'No history of anything,' a wise man once said, 'will ever include more than it leaves out.'[1] It is difficult to imagine a better exemplar of this insight than the history explored in this book. Seeing in some form or other has been around for hundreds of millions of years. It is a near universal but highly subjective experience among humans and across the animal kingdom. It is a complex neuro-physiological process that natural philosophers have struggled to understand for centuries, and its deeper workings are only just beginning to be unravelled. What's more, there are dozens of fascinating ways seeing can go wrong.

The definitive history of seeing may one day be written, but that day has not yet come.

In researching and writing this book I have picked a course through the millions of words written about the many different aspects of seeing. I've forged a path through a dense forest that traverses dozens of different fields of expertise. There is a logic to my path but others would inevitably have made different choices along the way. One way or another, a lot of territory remains unexplored. Despite my best efforts there are, no doubt, twists and turns and views I've failed to spot along the way, and the odd misstep. I apologise in advance for these, and welcome correction.

Susan Denham Wade

PROLOGUE: 2015

Early in 2015, Grace McGregor was looking forward to her wedding on the tiny island of Colonsay, two and a half hours by boat from the Scottish mainland. As the bride-to-be was making her plans, 300 miles away in Blackpool her mother, Cecilia, was shopping for her mother-of-the-bride outfit. Cecilia sent Grace pictures of several dresses she was thinking of buying, taken in a store on her partner's phone, then called her from the store.

Grace asked her mother which one she liked best.

'The third one,' said Cecilia.

'Oh you mean the white and gold one?' said Grace.

'No. It's blue and black,' said Cecilia.

'Mum, that's white and gold,' said Grace. When Cecilia insisted the dress was blue and black, Grace showed the picture to her fiancé, Keir. He agreed with Cecilia that the dress was blue and black. Keir's father was called in from next door to give an opinion. He thought it was white and gold.

The debate continued and spread to the couple's friends and family. Some people saw the dress in Cecilia's picture as blue and black, some saw white and gold. After a few weeks of local arguments, a friend of the couple called Caitlin McNeill put the photo on the social media website Tumblr and asked her followers to 'please help me – is the dress white and gold, or blue and black? Me and my friends can't agree ...'[1]

Within half an hour the picture found its way onto Twitter and with that became a hashtag: *#thedress*. It spread around the web like wildfire. Buzzfeed picked it up and asked its users to vote for white and gold, or blue and black. Now the Twittersphere erupted. At its peak, the hashtag was tweeted more than 11,000 times per minute. Eleven million tweets in total were posted overnight. Comments came from reality TV star Kim Kardashian (white and gold), who disagreed with her husband, Kanye West (blue and black). Pop stars Justin Bieber and Taylor Swift also saw blue and black, the latter tweeting that she felt 'confused and scared' by the phenomenon.

The next morning the picture featured on television news reports around the world, with newscasters arguing on air about the colours of the dress. Not only could no one agree, they couldn't comprehend how anyone else could see it differently from themselves. Even when they were told the dress was blue and black people couldn't change the way they saw the image.

As the debate continued, the media tracked down the family behind the original photograph. The weekend before *#thedress* went viral Grace and Kier had had their wedding as planned and gone off on honeymoon. The Ellen Degeneres chat show persuaded them to cut their holiday short and flew the whole party to the US to tell their story live on air. At the opening of the show Ellen showed the studio audience the original image – with which they were clearly already familiar – and asked them to indicate whether they saw blue and black, or white and gold. Sure enough, they were split. Later in the show she brought on Grace and Kier and their friend, Caitlin. They told their story on air, and the couple were rewarded with another honeymoon, this time in the Caribbean, and $10,000 cash to 'start their new lives'.

The climax of the interview came when Ellen called Grace's mother, Cecilia, onto the set. She walked on stage to cheers and applause, wearing the world's most famous dress. It was, unmistakably, blue and black.

BECOMING

HOW WE SEE

1

YOU SEE TOMAYTO, I SEE TOMARTO: THE SUBJECTIVE ART OF SEEING

Every man takes the limits of his own field
of vision for the limits of the world.

Arthur Schopenhauer, *Studies in Pessimism*, 1851

Had I been born 500 years earlier I would be blind. In their natural state my eyes see a world with no lines. Shapes are smudges and faces are blank. Colours merge into a murky brown and distances collapse to a single plane a few feet away. Everything is a complete blur.

But I was lucky enough to be born in the twentieth century. From the age of 8 I wore glasses and from 14 contact lenses – life-changing medical interventions so familiar they're hardly even thought of as technology – and my extreme myopia was corrected. As long as I had my specs on or my contacts in, I could live the same life as someone with 20/20 vision.

A few years ago I started fretting about my poor eyesight. What if there was a fire in the night and I had to leave the house without grabbing my specs? What if I got stuck somewhere for days with no glasses or spare contacts? I would be utterly helpless. It was a silly fear perhaps, but real; the universe of possible disasters expands as we get older, I've noticed. In any case, after more than thirty years I was tired of wearing glasses and fiddling around with contact lenses. Every week, it seemed, someone else regaled me with their successful laser surgery story. The time seemed right to take the plunge myself.

It turned out I was too blind for laser surgery, but I was eligible for lens replacement. That's the Clockwork Orange procedure where you sit in a chair with your eyes clamped open while the surgeon mashes up your lens, plucks it out, and replaces it with an artificial one that corrects the faulty sight. Thousands of people undergo the same procedure every day to treat cataracts. After a couple of mishaps and a bit more painless visual torture it worked. Now, for the first time in my life, when I wake up in the morning I can see exactly what the next person can see.

But I don't. And neither do you.

Looking Alike?

In Western societies more than 99 per cent of people share the daily experience of sight.* We can look at the objects around us and describe them using the same words: a red apple, a white cup, a wooden chair. We can recognise each other when we meet and translate marks on a page into language. The shared

---

* World Health Organisation figures state there are 36 million blind people in the world, and 217 million with moderate to severe vision impairment, based overwhelmingly in low income countries. (*WHO Fact Sheet #213*, accessed at www.who.int/en/news-room/fact-sheets/detail/blindness-and-visual-impairment.)

experience of our visual world seems complete: the world is as we all see it, together.

But the sense of commonality about what we see is an illusion. While two people may have identical visual capability, and so *can* see the same, no two people *do* see exactly the same. Every aspect of visual perception is subjective, unique to the perceiver. It isn't just beauty that is in the eye of the beholder, it is every single thing we see.

This subjectivity is down to the way seeing works. Human and all other vertebrate eyes are called 'simple' eyes as they have only one lens (insects and other arthropods have 'compound' eyes with many lenses). Simple eyes are structured superficially like a camera. Light comes in through a small opening (the pupil) and a lens focuses it onto a light-sensitive area at the rear of the eyeball (the retina), just as a camera lens focuses light coming in through the aperture onto a film. That's where the analogy ends, however. Unlike a camera, eyes don't capture the image in front of them then send it off 'upstairs' to the brain to be developed, like a film going off for processing. Vision is a pathway, an information processing system,[1] from the way the eyes gather visual information to analysing its components, to building up a conscious perception of sight and recognising the scene being observed. Different parts of the brain are involved at each stage, bringing each individual's experience, memory, expectations, goals and desires to bear on everything they see – or don't see. Neuroscientists call gathering and processing light signals – the physical seeing, if you like – 'bottom-up' processing, and the mechanisms the brain uses to influence vision – turning seeing into perceiving – 'top-down' processing. It's only in the last few decades that they have begun to understand the interplay between them.

The brain's involvement in seeing starts before we even see anything, with the way the eye gathers visual information.

While cameras capture an entire image in one shot, eyes don't. A photographic film has light-sensitive chemicals spread evenly across it, so the whole surface of the film reacts equally and immediately with the light that comes in through the aperture. The retina is very different. It is an outpost of the brain, formed in the early weeks of pregnancy from the same neural tissues as the embryonic brain and covered in neurons. Two types of photoreceptors – rods and cones – detect light and turn it into electric signals. They do quite different things and are spread very unevenly across the retina.

Cones can detect colours and provide excellent visual acuity but need relatively bright light. We use them for high-resolution daytime (or artificially lit) vision. Most of the eye's 6 million cones are concentrated in a tiny area in the centre of the retina called the fovea, the eye's central point of focus. They run out quite quickly away from the focal point.

Rods are roughly a thousand times more light sensitive than cones, capable of seeing a single photon, the smallest unit of light. They are extremely good at detecting motion, but they cannot see colours and they provide relatively poor resolution. That is why our night vision is colour-blind and relatively blurry. There are around twenty times as many rods as cones, clustered in the mid to inner part of the retina, outside the fovea, and continuing out in gradually decreasing density to the retina's edge.* Rods provide our night and peripheral vision.

The concentration of cones in the tiny fovea means eyes can only focus on a very small area at a time. At an arm's length

---

* You can see the difference between rods and cones by holding a coloured object at arm's length in front of you. Holding your gaze to the front, slowly move the object around to the side, wiggling it as you go. Quite quickly you will no longer see the object's colour, and it will become very blurry, but you will continue to be aware of movement even when you can't see what's causing it.

from the eye, the zone of sharp focus is only about the size of a postage stamp. Test this for yourself by holding up this book with your arm outstretched and looking at a single word. All the words around it will be blurred. We compensate for this tiny focal area by constantly and very rapidly moving our eyes around whatever we are looking at, three or four times per second in tiny subconscious movements called saccades, gathering more and more detailed information.

Eyes don't move like a printer scanning a document section by section. They move around a scene in all directions, fixing momentarily on something, then moving on, piecing an image together one postage stamp at a time. In the 1960s a Russian psychologist called Alfred Yarbus devised an evil-looking apparatus with suction cups like giant contact lenses that he placed over his subject's eyes while a camera tracked and recorded where they moved as they looked at various images. He traced the recorded movements onto the images they were looking at, showing the course of the eyes' journey and where they paused.

Yarbus discovered several extremely important things about vision. Firstly, saccades aren't systematic, but nor are they random. Eyes don't attempt to get around the entire scene being observed, but seek out the information that is most useful. From a biological point of view the most 'useful' information is what helps us survive. Thus, as Yarbus demonstrated, eyes are drawn towards images of other living creatures, especially humans, and particularly the face, eyes and mouth. These are the most important body parts for survival because they reveal important information about a person's intention and mood.

Yarbus also discovered that, when looking at a scene, our eyes try to interpret it in a narrative way, to piece together a story that tells us what is going on. Eyes move back and forward from one character to another, and to details in the scene the brain thinks will be important in understanding what is happening.

Somehow our eyes and brain stitch all this together into a coherent impression of what we are looking at, ignoring the movements in between each eye fix. It's similar to the way a film editor works, cutting together various shots to guide the audience through a scene's story. Incidentally, film editors have learned that editing cuts are most pleasing to the audience if they are made during motion. Harvard neuroscientist Margaret Livingstone believes this is because our visual system is accustomed to processing a series of shifting scenes (eye fixes) separated by movement.[2]

Later eye movement studies reveal that a person's cultural background can influence their gaze patterns. In one experiment two groups, one Western and one East Asian, were shown a series of images of a central object set against a background scene, such as a tiger in a forest or a plane flying over a mountain range. The Western group tended to focus on the main object, while the East Asian group tended to shift their gaze between the main object and the background.[3] The researchers proposed the reason for the difference was that Western culture values individuality and independence, hence the focus on the central character, while East Asian cultures are more interdependent, and thus those subjects were more interested in the context within which the central object was placed.

More recent neurological studies have shed light on this early stage of visual information gathering and how it contributes to vision. Light signals from the retina travel to two places – the thalamus, of which more in a minute, and the superior colliculus. The superior colliculus, which also has a major role in controlling head and eye movements, combines light signals from the retina with input from other parts of the brain – including areas responsible for memory and intention – to determine where the eye looks next. Have you ever looked around suddenly, but not been quite sure why? This was probably because the rods in your peripheral vision unconsciously detected some sort of movement, and your top-down system, realising this might mean danger, directed your eyes to examine the situation more closely. This is a basic survival response.[4] Thus from the very outset seeing is a combination of eye and brain, whether we realise it or not.

The top-down brain is also pivotal in filling in the parts of the scene that the eyes don't actively focus on. The rods and cones outside the foveal area provide a rough visual indication of the scene surrounding the focal point, and the brain fills in the rest from memory and experience. This gives us the confident – but quite erroneous – impression that we've seen the entire scene.

Sometimes our brains direct our eyes to focus on the wrong things, leading us to miss important information. This is the stock in trade of magicians, fairground tricksters and pickpockets. They are all experts at getting us to focus on irrelevant details while they deceive us before our very eyes. Even the apparently unmissable can become invisible when we are focusing on something else. In 1999, a Harvard research team showed subjects a video of a group of people throwing a ball and asked them to count the number of passes made. Half of the subjects didn't notice a gorilla walking right across the court as they were watching.[5] Similarly, we often miss quite major changes to what we are seeing. The same Harvard team sent a researcher posing as a tourist into a park to ask a passer-by for directions. As the researcher/tourist and the passer-by talked, two other team members walked between them carrying a door and swapped the 'tourist' for another researcher. Most of the passers-by didn't notice the change and carried on talking to the second researcher.

As light signals are received from the fovea and the rest of the retina they travel to the thalamus and are relayed from there to the visual cortex located at the back of the brain. This first stage of processing deals with basic visual signals such as whether a line is horizontal, vertical or diagonal, and was discovered in the 1950s by physiologists David Hubel and Torsten Weisel. In a ground-breaking study, they inserted microscopic electrodes into individual cells within the visual cortex of a cat's brain. They immobilised the cat with its eyes trained onto a screen and attempted to record its brain's responses to different light patterns. Over several days they shone lights all over the screen but couldn't get any of the cat's brain cells to respond. Eventually they tried a glass slide with a small paper dot stuck on it. As they moved the slide around they finally got a response: a single cell in the cat's brain started firing. They continued moving the slide around, trying to pin down where on the screen the dot set off the active brain cell. After many puzzled hours they realised it wasn't the dot that was causing the cell to respond, but the diagonal shadow cast by the edge of the slide when it moved across the face of the projector.

This was a completely unexpected result. After many more experiments the pair concluded that within a part of the visual cortex (now known as V1) each of the millions of cells is pro- grammed to respond to a single, very specific visual feature. A different cell responds to each of /, \, −, | , and so on. From these basic signals the brain can quickly build an outline of a scene – effectively a line drawing of whatever the eye is looking at. This is why we naturally recognise simple line drawings: they replicate the most basic way the brain processes images.

Hubel and Weisel's experiments were revolutionary because they showed that the brain doesn't actively analyse visual information. On the contrary, it reacts. Each individual cell within V1 either fires or doesn't fire automatically in response to the visual properties of a particular light signal. In the next stage of processing, V2, specific cells respond to contours, textures and location. Once again depending on the visual characteristics of each object – in this case, say, colour, shape, or movement – certain cells do or don't fire. Perception is formed by the combination of all the cells that fire in response to an image's various visual properties. This was an extraordinary conclusion and entirely contrary to what researchers had assumed up to that point. Hubel and Weisel later won the Nobel Prize (1981) for this work (though no prizes for kindness to cats) and their insights have underpinned research into the workings of the visual system since.

From the visual cortex, information is relayed through one of two pathways – the 'Where' pathway, common to all mammals, and the 'What' system that we share with only a few species. The 'Where' pathway is located in the parietal lobe at the top of the brain towards the back. It is colour blind but detects motion and depth, separates objects from their background, and places things in space. These are the basic aspects of vision required for survival, as they enable the seer to detect possible food sources and danger, and to move within their environment.

The second pathway, the 'What' system, takes place in the temporal lobe at the sides of the brain, over the ears. This pathway perceives colours and, critically, recognises things. It is a more sophisticated system than the 'Where' system and is thought to be present only in humans and other primates and, possibly, dogs.[6]

The human 'What' system has a particular region in the brain dedicated to recognising faces – a function of our deep history as a social species in which faces are extremely important to our survival (and why our focus is drawn to faces, as we've seen). That is why a young child's drawings of a person are almost always of a face with stick arms and legs: the face is instinctively what is most important. When we look at faces our brains compare the features we see with a stored database of 'average' facial features – eye width, length of face, nose size and so on – all within a split second. Caricaturists exploit this to create pictures that exaggerate the facial features that differ most from the norm. We recognise these images instantly because the artist is deliberately doing the same thing our brain does unconsciously.[7]

As we saw with the Yarbus experiments, the objective of visual processing is to understand what is going on around us, rather than to establish an accurate optical representation. Some top-down mechanisms add information to an image, wh ch i w y yo c n rea his se t nce. Others take away extraneous detail or adjust what we see to compensate for ambiguity. These measures allow us to survive and thrive but also leave us vulnerable to a wide variety of optical errors and illusions.* Many of these are so powerful that, even when we know what the illusion is, it is impossible to 'see' the optical reality. Consider a chessboard in partial shadow. In terms of its optical properties, a dark square in bright light might be lighter than a light square in shadow. Nevertheless, our eyes will always see a darkened light square as lighter than a brightly lit dark square. Our top-down system is using our past experience and memory to direct our perception here. It is an interesting philosophical question as to which version of the chessboard represents the 'truth'.

Bearing in mind the complexity of the visual processing system, and the varying role the brain plays at every stage of it, one may well imagine that people of different backgrounds might see the same thing differently. Recent research has uncovered several examples of significant perceptual differences across groups.

---

* German musician and visual artist Michael Bach has a wonderful set of optical illusions online at www.michaelbach.de/ot and see a short video of optical illusions at www.youtube.com/watch?time_continue=76&v=z9Sen1HTu5o.

A 2016 study demonstrated that obese people perceive distances differently from people of average weight. When asked to judge a 25m distance, a 150kg person estimated its distance as 30m, while a 60kg person judged the same distance to be 15m. The researchers put this down to a link between a person's perception and their ability to act – the assumption being an obese person would find it more difficult than a slim person to travel the same distance.[8]

The Himba tribe in Namibia continue to live a traditional life away from Western influences. Their language describes colours completely differently from ours. One colour, called *Dambu*, includes a variety of what we describe as greens, reds, beige and yellows (they describe white people as *Dambu*). Another colour, *Zuzu*, describes most dark colours, including black, dark red, dark purple, dark green and dark blue. A third, *Buru*, includes various blues and greens. Within their language and colour system, what we would call different shades of the single colour green might belong to three different colour families.

In 2006 researchers put this to the test.[9] They showed Himba people a set of twelve tiles, of which eleven were the same colour and one was different. In the first test the tiles were all green, with one being a slightly different shade. To most Westerners the tiles looked identical, but the Himba volunteers spotted the odd one out immediately. The second experiment showed the volunteers eleven identical green tiles and one that, to Western eyes, immediately stood out as being blue. The Himba, however, had difficulty differentiating this tile from the others.

Do You See What I See?

Around the time of my eye surgery *#thedress* happened. At a time when vision was uppermost in my mind, I started wondering: if it is so easy for people living in a similar time and place to see things differently, how different must the world have looked to people living hundreds or even thousands of years ago?

This was never going to be a easy question to answer. Nevertheless, I began digging around looking for clues. I read books and academic articles, visited museums, galleries and ancient places, and talked to experts. While I couldn't see through the eyes of ancient peoples, I discovered a lot of things that surprised and intrigued me. The more I found out, the more intrigued I became.

Eyes, I discovered, have existed longer than any other part of our body. Their structure has remained virtually unchanged through most of evolutionary history, even while the heads and bodies that housed them changed dramatically. Our eyes are almost identical to those of the very earliest vertebrates – our ancestors – eel-like creatures who lived in the sea more than 500 million years ago (mya).

But the most primitive eyes go back 100 million years further than that, back to the time when every living thing on Earth was still microscopic, until something triggered an explosion of frenzied evolution that resulted in the earliest animal kingdom. What sparked that explosion isn't certain, but a good candidate for the trigger is those primitive eyes. Millions of years before anyone coined the term, eyes may have been the original super-disruptor.

When the hominids – the immediate ancestors of humans – came along, they weren't content with their natural vision, venerable or not. They mastered fire, the first disruptive technology, giving them precious light through the night for hundreds of thousands of years, and changing humankind's place in the ecosystem forever.

Many thousands of years later, their descendants started making images of the world around them – pictures. Art was born. A few millennia after that they discovered how to polish glass into a mirror and see themselves reflected back. Then, just a few centuries further on, someone invented the first writing system, which eventually captured spoken language in a visual form. Writing was the beginning of what we call history and enabled the world's first civilisations to develop.

Centuries later, an Italian artisan ground two glass discs into lenses, joined them together in a frame and made spectacles. Two hundred years after that, a German goldsmith invented the printing press and spread literacy and learning throughout the known world, with dramatic consequences. A century and half later a Dutchman put two lenses in a tube and created a telescope, tilting the world on its intellectual axis. Many scientific advances and a few more centuries on, light was released from the bounds of the hearth and the wick and rechannelled into pipes that lit city streets, homes and the new factories like never before. The modern world as we know it had arrived.

In the nineteenth-century spirit of active enquiry and enterprise, two amateur scientists invented different versions of photography within weeks of one another. By the end of that century, still images had become motion pictures, which eventually came into homes as television.

Then just over a decade ago, a charismatic Californian entrepreneur launched the first smartphone. We've had our eyes glued to glowing screens ever since.

Each of these eleven inventions – firelight, pictures, mirrors, writing, spectacles, the printing press, telescopes, industrialised light, photography, motion pictures and smartphones – changed the way people saw the world. But each of them also changed the world into which they came: some immediately and with great fanfare, others more slowly and subtly but, I argue, no less dramatically. With each new visual technology the world was seen differently and became a different world. And with each new invention, vision slowly eclipsed our other senses, eventually relegating them to supporting roles in pleasure and leisure.

A dozen years into the smartphone era, it's time to take a look back at previous epoch-defining visual discoveries and ask ourselves the question: have we gone as far as the eye can see?

2

PERFECT AND COMPLEX:
EYES IN EVOLUTION

To suppose that the eye with all its inimitable contrivances for
adjusting the focus to different distances, for admitting different
amounts of light, and for the correction of spherical and
chromatic aberration, could have been formed by natural
selection, seems, I confess, absurd in the highest degree …
[But] The difficulty of believing that a perfect and complex
eye could be formed by natural selection, though insuperable
by our imagination, should not be considered
subversive of the theory.

Charles Darwin, *On the Origin of Species*, 1859[1]

In the beginning, there was light, but nothing was there to see it.

For several hundred million years after a giant cloud of gas
and dust formed our solar system, the Earth was a molten mass
of turbulence. Eventually the surface solidified into great, grey
continents peppered with rumbling volcanoes and surrounded
by vast reaches of ocean. The Sun shone by day and the Moon
by night.

There was light and dark, lightning and rain, rocks and water, but it all went by unnoticed.

One day, about 3.8 billion years ago, the first life appeared. No one really knows how it happened. Tiny, single-celled organisms lived underwater, drifting around in the warm, chemical-rich currents rising below the surface of the young planet's huge seas.

Life moved very, very slowly for about 3 billion years. Gradually the single-celled organisms evolved into different forms that would eventually become the major kingdoms of living things: bacteria, algae, fungi, plants, and animals. Around 650 mya some organisms developed a bilateral body form, with a front and back, and left and right sides. They were still too small to see and, aside from a microscopic mouth and anus, were featureless, worm-like blobs. It was only millions of years later when experts examined their fossilised forms under powerful microscopes that any of these changes were distinguishable. The bilateral group split further into what would eventually become the vertebrates (mammals, birds, fish, and reptiles) and the arthropods (insects, spiders and crustaceans) about 600 mya, but all remained invisible to the naked eye.

Then about 540 mya, something changed. After 3 billion years of glacial change the microscopic animal kingdom exploded into a frenzy of evolution. Creatures morphed, died out, grew, proliferated, changed, died out, mutated, reproduced, proliferated, died out, reproduced, mutated, thrived, grew features, reproduced, grew, died out, mutated and proliferated, again and again, generation after generation.

After about 20 million years of frenetic change, everything slowed down again. Winning species emerged and relatively stable ecosystems were established. But the underwater world was transformed. Murkily invisible swarms of microscopic life were replaced with complex, visible flora and fauna, a diverse range of species both large and small. They had characteristic shapes and features including limbs, teeth, antennae, gills, shells, spines and claws. Some had optical features like stripes and iridescent colours.

At the top of the food chain was the *Anomalorcaris*, a large predator with a soft, segmented body shaped a bit like a stingray. It had a fish-like tail and a pair of hooked, grasping appendages projecting from its head, between two bulging eyes, and it could grow up to 6ft long. There were many varieties of *Trilobite*, giant woodlice-type creatures with long antennae and a pair of compound eyes on the top of the head. *Opabina* was a bit like a prawn with five eyes and a long nozzle appendage at the front. The predator *Nectocaris* vaguely resembled a squid and had eyes on short stalks.[2] They hunted and were hunted, scavenged, reproduced, lived and died under the sea for tens of millions of years until an extinction event wiped many of them out around 488 mya.[3]

Palaeontologists call the period of rapid evolution – known as adaptive radiation – that happened 540 mya the Cambrian Explosion. It is probably the most important evolutionary event in the history of life on Earth. It is also one of the biggest mysteries. Before the Cambrian Explosion life forms comprised undistinguished, microscopic organisms. Just 20 million years later there was a richly diverse, and visible, underwater ecosystem. No one knows exactly what happened in the intervening years.

The animals that appeared during the Cambrian Explosion are documented in great detail by modern palaeontologists thanks to some catastrophic events that rocked the underwater world more than half a billion years ago. Very occasionally, a huge expanse of mountainside slid suddenly down a steep underwater slope onto the seabed below, engulfing all the surrounding plants and animals. These particular mudslides were formed of exceptionally fine sediment that immediately made its way into every nook and cranny of the unsuspecting wildlife below. The events happened so suddenly and with such force that all the oxygen and bacteria present in the ecosystem were evacuated, leaving the captured organisms vacuum-packed in the exquisitely insidious mud.

The trapped organisms turned into extraordinarily complete and detailed fossils – almost perfect three-dimensional snapshots of the scene at the moment of devastation. Unlike most fossils that form gradually as silt builds up over a decaying organism, these were formed almost instantaneously, capturing soft tissues that usually rot away long before fossilisation takes place, sometimes down to microscopic details. They revealed species and body parts that almost never make it into the fossil record, including eyes.

The aquatic communities these extraordinary mudslides captured were frozen in time – Pompeiis of some of the earliest visible life on Earth – until just over a century ago. In the summer of 1909, self-taught palaeontologist and geologist Charles Doolittle Walcott took his family on a fossil-hunting holiday in the Canadian Rockies. They collected a large haul of fossils including species they had never seen before, and 'several slabs of rock to break up at home'.[4]

Walcott had stumbled upon a *Konservat-Lagerstätte* (from the German for 'conserving storage place'), an equatorial marine-scape frozen in time around 508 mya, just a few million years after the Cambrian Explosion. Over eons the Earth had shifted, moving the ocean bed from the equator to western Canada. The Burgess Shale, as the fossil field that Walcott discovered is now called, revealed more than 150 new species and more than 200,000 extraordinarily well-preserved specimens.

In the century since Walcott's find, palaeontologists have uncovered other Cambrian Era *Lagerstätten* in Australia, China,

Greenland and Utah, USA. They reveal that many of the Cambrian animals had eyes. There were creatures with two eyes, four, and more. Some were on stalks, some at the front of the head, others at the back or the side, some scattered over the body. Eyes varied but they were not unusual. Analysis of particularly well-preserved fossils has revealed that compound eyes from as long ago as 520 million years had vision equivalent to that of a modern insect. Researchers at the *Lagerstätte* in Chengjiang, China, have further established that vision among the earliest animals was closely associated with a predatory lifestyle. They found that around a third of the specimens discovered had eyes, and of those with eyes, 95 per cent were mobile hunters or scavengers.[5]

The Light Switch Theory

The very early appearance of functioning eyes and the association of eyes with predatory behaviour prompted a theory that the development of primitive vision may have triggered the Cambrian Explosion. Oxford zoologist Andrew Parker's 'Light Switch' theory maintains that, as tiny creatures slowly became more complex, some started to develop a form of vision.[6] It began with a light-sensitive spot somewhere on the body then mutated further over generations until it became a primitive eye.

It is easy to imagine the survival advantage animals with even a primitive form of vision would have. They could avoid danger, and instead of floating about passively waiting for food to happen by, could seek it out.

Once creatures started actively searching for food, some inevitably became predators and others, correspondingly, prey. It was the opening line of the story of the animal kingdom. But vision alone was not enough. Remember that body forms at the outset were tiny and indistinct. To become active in seeking food, predators needed mobility and weapons. Prey needed defences and means of escape. Sure enough, evolution's random process of mutation and proliferation gradually provided these. Predators developed claws, teeth, appendages for swimming. Prey species formed hard shells, spines, means to burrow into the sand, even camouflaging colours.

New traits evolved rapidly to out-compete other species, who responded with further adaptations. It was an evolutionary arms race between predators and prey as every species had to adapt or die.

Eventually some became new species with varied body parts, different ways of moving through the water, and diverse feeding strategies. None of this was conscious, it was all trial and error: millions of organisms reproducing, mutating, dying out, mutating, thriving, over and over again, generation by generation. It didn't come about overnight, but in evolutionary terms it happened very fast. In the face of a rapidly changing environment, the advantage or disadvantage of a random mutation becomes apparent very quickly: creatures either survive or are quickly wiped out. When we're talking in terms of millions of years, small changes can become major changes in a relatively short period of time.

To take the specific example of an eye, in 2004 a Danish zoologist built a theoretical model to estimate how quickly a light-sensitive patch on an animal's skin could evolve into a sharply focused camera-type eye like our own. The model assumed a small improvement in each generation. It calculated that the entire transformation from patch to eye could take place in less than half a million generations.[7] This would take 10 billion years for humans, but for a creature with a life span of months or even days, it could take no time at all.

If vision triggered the Cambrian Explosion, as the Light Switch theory maintains, it would have to exist *before* the Explosion began. But all the fossil evidence for species with eyes comes from *after* it settled down.

There is another way to probe the deep past, however. All living things carry with them their entire evolutionary history, locked into the genes that reside in every cell. Rapid progress in genetics over the last few decades has managed to unpick that lock, allowing zoologists to reach back through time and find connections between species that physical observation can miss. Every species has thousands of different genes, each made up of complicated sequences of DNA. If different species share a matching DNA sequence, then almost certainly those species share a common ancestor, because the chance that they would both evolve the same version of something as complex as a gene independently is so small as to be virtually impossible.

In 1994, a Swiss biologist called Walter Gehring wanted to test his hypothesis that the Pax6 gene, found in every modern animal species, prompts eye growth. In a laboratory experiment he inserted Pax6 from a mouse into fruit fly embryos, in different places corresponding to particular parts of the adult fly's body.[8] The fly embryos hatched with an eye growing where the Pax6 gene had been inserted. The fly with Pax6 inserted into its antenna area grew an eye on its antenna. Another grew an eye on its leg, while another still grew one on its wing, all corresponding to where the gene had been inserted.[9] But crucially, although the introduced Pax6 came from a mouse, the extra eyes weren't mouse-type eyes. They were compound fly eyes, fully formed and responsive to light. Gehring concluded Pax6 does prompt eye growth, but doesn't determine eye type. Eye type genes must have evolved later, after the two animal groups had diverged.

This result tells us that the common ancestor of mice and flies had the eye growth gene Pax6, so must have had some sort of primitive eyes. Every eye that exists today evolved from these, the proto-eyes of a microscopic common ancestor living 600 mya when the ancestors of mice and flies diverged. The experiment established that eyes – and therefore vision – existed 60 million years *before* the Cambrian Explosion.[10]

Darwin wrote that the idea that eyes could have evolved through natural selection seemed 'absurd'. If the Light Switch theory is correct – and it seems to be supported by fossil and genetic evidence – eyes were not only the product of evolution, but the spark that prompted the evolution of life on Earth as we know it.[11]

If so, seeing wasn't the outcome of the biggest evolutionary event in history – it was the cause. The advent of vision turned 3 billion years of settled life on its head and caused 20 million years of evolutionary chaos. Hundreds of millions of years before human technology appeared, seeing itself was the original super-disrupter.

Eternal Eyes

What about our own eyes? What has happened to eyes in the eons since they first appeared? How have they changed? What makes human eyes special? The answer to all these questions is: not much.

Humans have simple (single-lensed), camera-type eyes. Chinese palaeontologists have found fossil remains of 520 million-year-old simple eyes but they weren't complete enough for them to be sure whether they were camera-type eyes like our own, or to assess how well they could see. Again, we need to look beyond the fossil record to discover how and when human eyes evolved.

Like geneticists looking for a common gene, zoologists can trace the evolution of a particular complex trait by looking for species, living or extinct, with the same trait. If two species have a complex trait in common, and share a common ancestor, it is very likely that the common ancestor had the trait in question. In the same way that common genes indicate a historic relationship, the same complex trait is unlikely to have developed independently in two species after they diverged.

Australian neuroscientist Trevor Lamb has used this principle to trace the deep history of the human eye, working backwards up the evolutionary tree combining information from fossils, living species, and genes.[12]

The first steps up the human family tree are straightforward. The camera-type structure of human eyes is shared by all mammals. Birds, fish, reptiles and amphibians also have a camera-type eye structure. The last common ancestor of this group – the jawed vertebrates – lived around 420 mya, which means that camera-type eyes are at least this old.

The next step back takes us to an eel-like creature called a lamprey. The lamprey is an ancient and primitive creature; there are fossilised remains of lampreys from 360 mya[13] and they are thought to have hardly changed in over 500 million years.[14] It has a primitive cartilaginous skeleton, making it a vertebrate, and as such the species is of great interest to biologists studying the early evolution of vertebrates. What differentiates the lamprey from other vertebrates is that it has no jaw. Instead it has a permanently open ring of a mouth filled with spiky-looking teeth that latch on to a victim's flesh so the lamprey can suck its blood. Lamprey was a delicacy in the Middle Ages, and King Henry I famously died after eating a 'surfeit of lampreys' in 1135.

Lampreys are not pretty – 'killer lampreys' featured in a recent horror movie[15] – but for our purposes they have one feature of particular interest. Lampreys have a camera-type eye structure essentially the same as our own, with a cornea, lens, iris and eye muscles, a similar three-layered retina and photoreceptors like our rods, although they don't have cones. This makes it extremely likely that our common early vertebrate ancestor also had camera-type eyes when the vertebrates diverged into jawed and jawless species around 500 mya.[16]

The trail ends here. The next step up the evolutionary family tree takes us back to a microscopic ancestor that lived 100 million years earlier, before the Cambrian Explosion. The living descendant of this creature only has a simple eye spot, indicating that the common ancestor had at best a rudimentary eye. That leaves us with the lamprey as our oldest seeing relative, giving our camera-type eyes a highly respectable vintage of at least 500 million years.

In half a million millennia camera-type eyes changed very little. Essentially the same eyes served all the descendants of our common ancestor including fish, dinosaurs, birds, mammals, reptiles and humans … and lampreys.

However, the bodies that hosted those eyes changed a lot, giving similarly structured eyes quite different capabilities. Carnivorous dinosaurs such as *Tyrannosaurus rex* were the first to develop heads with forward-facing eyes. This gave them binocular, three-dimensional vision, valuable for hunting and chasing prey.

Predators that are low to the ground and hunt by ambushing, such as crocodiles, snakes, and some small cats, have pupils with vertical slits that fine-tune their distance perception so they can judge the moment of attack precisely.

Prey species, meanwhile, evolved visual traits to help them evade capture. Hunted animals need to maximise their field of vision to detect potential predators. Many herbivores evolved protruding eyes on the sides of their heads, like rabbits, and have almost 360-degree vision. Some grazing animals including sheep, goats and antelopes developed horizontal postbox-shaped pupils, to further improve their field of vision, and eyes that swivel when they bend down to graze, to keep their pupils horizontal and on the lookout.[17] Nocturnal animals such as tarsiers, the small googly-eyed primates, have very large eyes that improve their night vision. Tarsiers' eyes are bigger than their brains. Unfortunately for them (and other nocturnal prey), various predators have also developed night vision, most famously owls.

Human Eyes

When our ancestors, the primates, first appeared about 55 mya they were largely nocturnal omnivores, feeding on insects and small vertebrates. As predators, their eyes were forward facing and were large relative to their bodies to aid night-time vision. They may have been able to see ultraviolet rays.[18] When the apes and monkeys (our ancestors) separated from the other primates, they shifted to a daytime hunting pattern and their eyes evolved a higher concentration of light receptors in the fovea, giving sharper daytime vision, and a wider range of colour perception.[19]

When human ancestors diverged from the other apes, the appearance of their eyes changed. Ape eye sockets are round and dark, similar in colour to the surrounding face, and their 'whites' are also dark. Human eyes, by contrast, stand out clearly from the face. Iris and pupil are surrounded by a white sclera that is highlighted by the almond shape of the human eye socket.

This change makes it very easy to follow a person's gaze, and we are programmed from birth to do this. Human babies stare intently into their carer's eyes during feeding (to an extent that can be both uplifting and unnerving), and follow a carer's gaze, even when the carer's head doesn't move, from a very young age. They don't follow head movements if the carer's eyes are closed, implying the eyes are the critical feature. Apes do exactly the opposite. They follow head movements with or without a shift in gaze, but they don't follow eye movements alone.[20]

Seeing where another individual is looking gives away important information about them: their intended direction of travel, or that they have seen something interesting, or an enemy approaching. In a society where there is an assumed level of trust, communicating by eye movement is highly efficient.

However, a highly visible gaze can be a liability in different circumstances. In a competitive social environment, like that of chimpanzees, an individual who sends signals through eye movements might give away valuable information, such as where the best food can be found. Visible eye movements are a disadvantage.

Michael Tomasello, the zoologist who conducted the gaze experiments, believes that visible eye movements are evidence of very distinctive human characteristics. Tomasello believes that as hominins (predecessors to modern man) diverged from the other apes over the past 2 million years, with a corresponding divergence of lifestyles, their eyes evolved to support a more cooperative way of living. The whites of human eyes, he believes, are evidence of long-standing close-range cooperation within social groups, possibly reaching back before the development of verbal language.[21]

Our eyes may be the windows to the soul, but they are also windows into our history. Their positioning, size and shape, capability and internal structure all tell us about how our ancestors lived, and who our earliest ancestors were.

Darwin called eyes a 'complex and perfect' organ. They are not perfect; they bear various scars of the bumpy evolutionary process they resulted from. However, their great longevity demonstrates they must be perfectly good enough, or else they – and we – would have been superseded at some point by a better model through the relentless forces of natural selection.

TRANSFORMING

THE VISUAL
TECHNOLOGIES
THAT BEGAT HISTORY

STOLEN FROM THE GODS:
FIRELIGHT

I am he that searched out the source of fire,
by stealth borne-off inclosed in a fennel-rod,
which has shown itself a teacher of every art to mortals.

Prometheus, from Aeschylus (525–426 BCE), *Prometheus Bound*,
trans. T. A. Buckley, 1897

In a story told by Kalahari Bushmen, at the beginning of time the People and the animals lived under the ground with the great god and creator, Kaang. The People and the animals could talk to each other and understand each other and they lived in peace. Although it was under the ground, it was always light and warm, and everyone had enough of everything. One day Kaang decided to build a world on the surface. First He created an enormous tree with many branches. Then He created all the features of the land: the mountains, valleys, forests, streams and lakes. Finally, He dug a deep hole from the base of the tree down to where the People and animals lived. He invited the first man and the first woman to come up and see the world he had made. The man and the woman came to the surface and saw the world and the Sun and the sky and they were very happy. All the other People and animals came up through the hole too and saw the new world.

Kaang said, 'This is the world I have created for you all to live in. I have just one rule: do not make fire, as this will unleash an Evil force.' He left and went to a secret place to watch them all.

At the end of the day the Sun set. It became dark and got cooler. Everyone was scared, especially the People. None of them had been in darkness before. The People could not see in the dark like the animals could, and they did not have fur to protect them from the cold. They became afraid of the animals and didn't trust them. They forgot Kaang's words and built a fire. Now the People could see each other's faces in the light of the flames and they felt safe and warm.

But when they turned around the People saw all the animals running away. The animals were frightened by the fire, and they fled into the caves and mountains. The People were sorry, and called after the animals, but they didn't reply. Kaang came out of his hiding place and said, 'You disobeyed me and built a fire. It is too late to be sorry; the Evil force has already been released.'

Ever since that day the animals and the People have feared each other and lost their ability to talk and live together.[1]

The terror the People in the creation story felt on confronting darkness is a universal theme. In spite of its nightly inevitability, humans have always feared the dark. Language is peppered with allusions to our dread of lightlessness: dark deeds, black looks, dim view, gloomy, obscure, shady, shadowy, drear. The word darkness is used to describe ignorance, anger, sorrow and death,[2] while the forces of darkness are wicked and evil.

Night is the natural conspirator of darkness, and by association the dwelling place of evil. In Greek mythology the gods Erebus (darkness) and Nyx (night) were husband and wife. They spawned the demigods Disease, Strife and Doom.[3] On the other side of the world, the Maori goddess Hine-nui-te-pō represented night and death and ruled the underworld. In Germanic cultures, night hosted the witching hour when monsters such as vampires, ghosts, werewolves and demons emerged, and when black magic was at its most potent.

Darkness was also the bedfellow of death in ancient cultures from Mesopotamia to Egypt to China to Europe. Almost all cultures have stories of an underworld, bathed in eternal darkness, representing death and housing evil spirits.

Light, on the other hand, is almost always associated with the good, and often the divine. We see the light, are illuminated and enlightened. The first chapter of Genesis proclaims: 'God said, Let there be light: and there was light. And God saw the light, that it was good: and God divided the light from the darkness.' Many other cultures also associate the creation of the world with light.

Most of the world's major religious icons are associated with light. Jesus said, 'I am the Light'; the prophet Mohammed is *noor*, the light of Allah personified; Buddha is the enlightened one; light represents the almighty in Hinduism. Religious rituals and festivals also often centre around light: Buddhist festivals are all on full Moon days; Diwali is a festival of light; Hanukkah requires the daily lighting of a candle; Christians light a votive candle as an offering with prayer.

Satan, on the other hand, is the Prince of Darkness.

Fear of the dark is not irrational. Humans are adapted for a daytime existence, and although our eyes can detect very tiny amounts of light, our visual acuity, or sharpness, is poor in the dark. Darkness is frightening even today for the simple fact that we can't see very well.

But fear of the dark is also primal, ingrained in our species for millions of years. Many large carnivores were nocturnal hunters, and our ancestors were ready prey for them if they were caught on the ground after dark. That's why the prehistoric great apes (hominids) and their pre-human successors (hominins) slept in trees for millions of years, nesting safely above marauding predators. Most apes still do.

At some point, however, our hominin ancestors left the nocturnal safety of the trees behind. Somewhere along the path of human evolution, one of our ancestors figured out that they could challenge the night. With this capability they could come down from the trees and live safely on the ground among the fiercest of beasts. Little did they know that in doing so they would change the course of their own destiny, and that of every other species on Earth. What began perhaps as a comforting protection from the very real perils of the dark would become the most powerful technology in human history. What they discovered was firelight.

The First Fires

In the Bushmen's story, fire separated the People from all the other animals. This is a recurring theme in indigenous myths about the origins of fire. In 1930, the anthropologist Sir James George Frazer studied the fire-origin myths of traditional people from all over the world.[4] He found some other remarkably consistent themes, even across far-removed geographical zones. One common story describes a thief, often characterised as a bird, who steals the jealously guarded fire from a selfish deity or monster. Another common myth was that fire was bestowed on mankind by a benevolent god in recognition of their favour with him. The Greeks' fire-origin myth combines these two themes: Prometheus, a minor god, felt sorry for the weak and naked humans. He stole fire from Zeus and gave it to mankind.

Charles Darwin said fire was man's greatest discovery after language,[5] but fire wasn't a human discovery. It is a natural process, a chemical reaction that occurs when oxygen and fuel are ignited. Elemental as fire may seem, the ancient Greeks were wrong to group it with air, earth and water as one of the fundamental elements. Air, earth and water are substances. They were present long before there was life on Earth. Fire didn't appear until millions of years later.

The young Earth was speckled with active volcanoes, and frequently hit by lightning, but still there was no fire. Volcanoes provide what look like spectacular pyrotechnical displays, but there is no fire in a volcano. The fire-like components of a volcanic eruption have their own names. The glowing fountains that look like flames are lava. What appears to be smoke is actually a sulphur-rich mixture of gases released from below the Earth's surface called fume. The sparks and cinders that shoot up into the air are volcanic spatter. And volcanic ash is not ash at all, but the pulverised volcanic dust tephra. None of these are fire.

A lightning spark can be hotter than the Sun, but lightning isn't fire either; it is an electrical discharge in the atmosphere. Both fire and lightning create heat and light, but they are very different phenomena. Lightning and volcanoes can both ignite fires, but only if the other raw materials of fire are present.

Fire needs three things: oxygen, fuel and ignition. Without all three things it cannot exist. And there was no fuel on Earth until plants migrated out of the oceans and colonised the land around 440 mya.[6] When plants died their remains became Earth's first combustible material. Until there were dead or dying plants on the ground, there was no fuel. No fuel, no fire.

This makes fire considerably younger than eyes. As we saw in the last chapter, eyes similar to our own were already swimming around on the heads of our fishy ancestors at least 500 mya, tens of millions of years before a bolt of lightning or a red-hot drop of lava ignited some dead plants and started Earth's first fire.

The Taming of the Flame

Humankind's relationship with fire is very long and very deep. Exactly how long and how deep is a matter of ongoing scholarly debate. It is generally agreed that our human ancestors, the hominins, were using fire well before our own species, *Homo sapiens*, appeared around 200,000 years ago. But it is also possible that hominins were associating with fire millions of years earlier than that. Recent scholarship proposes that fire was more than a mere associate of the hominins, it was a life partner.[7] Furthermore, access to fire may not merely have been useful to early humans, but instrumental in several of the critical changes that marked our evolution from hominin to human.

That is to say, mankind didn't make fire, but fire may well have made mankind.

The story begins where all the oldest hominin fossils sites have been found,[8] in East Africa. The fault line known as the East African Rift System runs from the Red Sea down the eastern side of the African continent to the northern plateaus of South Africa. Between 5 and 7 million years ago there was a period of intense geological activity along the Rift.[9] Vast tracts of land were pushed up into mountains by volcanic activity, while other areas subsided into deep valleys. What had for millions of years been flatlands became a complex topography of mountains, plateaus, lakes and canyons. The climate and vegetation changed dramatically as warm, moist monsoonal air coming from the west was stopped by the rising mountain ranges. Meanwhile, east of the Rift, tropical forests died out and gave way to cooler and more arid woods and grasslands.

With these dramatic environmental changes going on around them, every species had to adapt or die. It was another example of the sort of adaptive radiation we saw with the Cambrian Explosion. The whole region became a hotbed of evolution as new species and traits appeared and disappeared in the trial and error of natural selection. One of our primate ancestors developed the ability to walk upright, separating it forever from its ape cousins. While there are dozens of theories as to why this was successful, the most straightforward is that walking on two legs uses less energy – and therefore fewer precious calories – than walking on feet and knuckles. This became a significant advantage as jungles turned into plains and food sources became more widely dispersed, requiring longer journeys to find them.

The first upright walkers were small, about the size of a modern chimpanzee, with a chimp-sized brain and protruding facial features with a large jaw and teeth. While they could walk upright on the ground, they still had shoulder blades well adapted for climbing, and they spent a fair amount of their time up trees, especially at night when large predators were most active. They looked like Lucy, the famous 1970s reconstruction of the first woman, made from an unusually complete *australopithecine* fossil from around 3.2 mya, found at the northern extreme of the Rift.

The first upright strides brought our ancestors face to face with fire. Fire was an ever-present fact of life for the Lucies. As well as active volcanoes, lightning strikes were frequent – to this day central Africa is struck by lightning more often than anywhere else in the world.[10] The rainforests east of the rift were dying, making them tinderboxes waiting for a spark, while the drier, sparser vegetation of the newly forming plains also provided ample fuel. Fire was a major factor in the environmental changes happening in the region.

From their treetop beds, Lucy and company would often see the light of nighttime flames. Sometimes they were forced to leave their nests to flee fires and used the light they emitted to find their way to safety. They saw other animals, including large carnivores, actively avoiding smoke and fire, and occasionally had to run alongside them to escape moving wildfires. In the face of a fast-moving fire there was only one predator: fire was always at the top of the food chain.

Eventually some of the Lucies evolved into a new species, *habilis*. They still looked ape-like but their brains were twice as big as those of any modern apes. They used knives made from rock flakes to cut meat away from bones and remove soft tissue from animal carcasses.[11] The species survived for hundreds of thousands of years, probably alongside *australopithecines*, before some of them discovered something that changed their future dramatically.

During the time of the *habilis*, fire was still a major factor in the ecology of the Rift, and they would almost certainly have encountered the aftermath of fires while foraging. At some point they learned – as modern chimpanzees have – that fire sites could be rich sources of food. Various seed and nut pods are opened by fire, and insects and small animals emerge from smouldering wood, ready to be caught. Beetles and cockroaches seek out recent fires to lay their eggs, attracted by the dead and dying timber that will provide food and shelter for their woodworm and termite larvae.[12] Those *habilis* brave and careful enough to avoid getting burned could find easy pickings at fire sites.

As it happened, many of these fire-associated foodstuffs were highly nutritious. Small vertebrates and insects provided calorie-dense meat, some providing as much fat and protein as the equivalent weight of modern-day beef. These animal products were also much easier to chew and digest than the raw diet the *habilis* usually ate, so required both less time and energy to eat. This provided a double nutritional bonus, giving the most fire-savvy individuals a survival advantage.

At some point this fire-savvy group started actively seeking out smouldering fire sites for the feast they might provide. They acquired some basic fire management knowledge, such as how to recognise which parts of a smouldering log were safe to touch. Opportunistic fire foraging became part of the lifestyle of some these *habilis*.

One day, an adolescent *habilis* picked up a smouldering stick and waved it around in a jovial threat. Modern chimpanzees pick up and throw burning sticks in a similar way. But this *habilis* had a much larger brain than a chimpanzee's and had perhaps anticipated the reaction his behaviour would provoke. He realised the power of the stick's scorching hot end to frighten or hurt others and remembered it. Others saw what happened and remembered it too. The next time the group came upon a fire site, several of the bravest individuals picked up burning sticks and had a play fight. Another time they carried the firebrands away to show them off to others. Then they took them further away still. While they were carrying fire sticks around one day they came upon a predator, who looked threateningly at them. They waved the firebrand at the predator, who backed off and eventually slunk away.

They realised that fire was a powerful source of protection, and it became even more sought after. Over time some *habilis* found ways to conserve fire for longer, by adding leaves and sticks to a fire site. They eventually figured a way of taking turns to keep the fire going, one staying with the fire while others gathered fuel. With fire sticks in hand they could range further to find food, knowing they were safe from predators, even after dark.

Someone realised a fire stick was not only a good form of defence, it could also be used aggressively. By waving burning logs they could scare predators away from a carcass, leaving it free for the *habilis* to steal the meat, and get to the better parts of the kill that the more powerful carnivore would never normally leave behind.[13] A brandished fire stick moved its possessor temporarily up the food chain.

Once the *habilis* could keep the fire going there was no need to shelter in the trees as soon as dusk arrived. They could gather together by the fire's protective light, knowing predators would not venture near. With firelight to provide light and protection, daytime could be devoted to finding food and fuel, leaving the evening free for other activities such as repairing tools or spending important bonding time together.

Meat became a regular part of the diet for these *habilis*, hunted with spears or stolen from other carnivores. They knew from experience at fire sites that burned meat tasted better than raw, and some of them tried different ways to use fire to burn some of the meat. Eventually a rudimentary form of cooking started, and this provided a major survival boost.[14] The raw leaves, fruits and tubers that formed a large part of the primate diet were hard and tough, and required a serious effort to break down. Modern apes in the wild spend almost half their day chewing.[15] Cooking softens food and makes it much easier to chew, so chewing time for the cooking *habilis* reduced dramatically. Spending fewer hours chewing left more time in the day for other activities: gathering more food, making tools, and bonding with others in the group. Cooking also changed their food's chemical composition and made it more digestible, providing more absorbable calories than the same thing raw, meaning the cooks had more energy to use for other metabolically expensive functions, such as brain activity.

All these fire-related changes prompted striking physical adaptations. Guts shrank as the fire-savvy *habilis* started to rely on easily digested cooked food. Jaws and teeth also shrank as chewing demands were relaxed. The combination of better diet, smaller gut and more sophisticated hunting and foraging activities prompted their brains to grow.[16] Legs and shoulders adapted to a terrestrial life, free from the demands of climbing and swinging in trees. Eventually they became unrecognisable. They were a new species, and a new genus: *Homo erectus*, the first ape to look truly human. *Erectus* was much taller than *habilis*, with longer legs and shorter arms fully adapted for walking and running rather than climbing. The ribcage was compact and the waist slim as the abdomen housed a much smaller gut. The face was flatter with considerably smaller teeth and jaw, although the brow remained pronounced and the forehead sloped backwards. *Erectus'* brain was more than 50 per cent larger again than that of *habilis*, and more than double the size of *australopithecine*.[17]

It would be many thousands of years before fire-starting techniques were discovered. For now the priority for *erectus* was to keep the fire going. This required cooperation: fuel had to be gathered and the fire tended. Cooking also required self-discipline. Instead of eating food as they found it, those gathering the food had to ignore their hunger and bring food back to camp to share with those who tended the fire. Group members who didn't cooperate were forced out; fire and cooking were too precious for the group to tolerate misfits.

Firelight, once controlled, became one of the earliest human technologies. The day extended beyond sundown, because *erectus* had discovered a tool for seeing in the dark. They realised at some point that they didn't need to use precious daylight hours for cooking; this could be pushed into the evening. They didn't feel drowsy at sundown as the light from the campfire suppressed melatonin production and delayed sleepiness for a few hours.[18] As darkness closed in, *erectus* group members would gather around the fire anticipating their meal, enjoying the light and warmth, and trusting in the protection of the group. With the day's serious business of food and fuel gathering complete, evening became an important time for social bonding.[19] This is vital to group living as it creates the trust and mutual obligation necessary to sustain a communal existence.[20]

Apes bond by grooming one another, but the early humans developed another way to bind the group. Their brains were sophisticated enough to make and understand simple jokes. As they sat around the fire in the evenings, they would tease and joke together. The resulting chatter and laughter released endorphins in the same way that grooming does for apes. But laughter and jokes were far more efficient than grooming: the whole group derived pleasure from them whereas only a single recipient got pleasure from grooming.[21] A few hours after dark, everyone went to sleep around the fire, knowing they were safe from predators. Every so often someone got up to tend the fire during the night. There was no longer any reason to climb into the trees.

The cooperative behaviour developed to sustain the group's fire extended into other activities. Males started hunting in groups, and by working together managed to kill animals much larger and faster than themselves. They used wooden spears with tips hardened in fire and carried hand axes for cutting and digging. Females gathered fruit and berries, and looked after children and the camp, especially the fire, which was now established as the home base. Whenever the group moved they carefully carried their fire with them.

Evening life centred on the campfire. At some point, fires were bound by rocks or made in a pit. Future archaeologists would call these hearths; the word comes from the Old German word for home.[22] A pattern was established. Daytime was used for economically productive activities and night-time for imaginative, bonding pursuits. This pattern would endure for hundreds of thousands of years, until another form of illumination changed everything again.[23]

Communication was vocal, with much laughter, but probably not what we would recognise as language. Perhaps some repeated vocal patterns emerged, similar to what we might call poems or songs that the group recognised and to which it responded. These activities served to bind the group and shape the identity of its members.[24] Proto-language was another aural form of grooming.

The *Homo erectus* line lived for about 1.5 million years and ranged far and wide. Their large brains and control of fire helped them adapt to many different environments. Some made their way out of Africa and into Asia, reaching as far as China and Indonesia. Others turned west around the Mediterranean and ended up in Georgia and as far west as Spain.

A few of the migrating *Homo erectus* seem to have lost their fire en route. Lightning and volcanoes were much less frequent outside Africa, and so it was harder to find natural fires to replace a lost flame. Over time, the skills for keeping a flame alive may also have been lost. The fireless groups continued to hunt, and developed strong bodies and large brains, but they moved in smaller bands and used simpler tools. Without the regular campfire to gather around it was more difficult to sustain larger groups. Fire was used only sporadically, after fruitful lightning strikes. Some of this group died out, while others eventually became *Homo neanderthalensis*.[25] The neanderthals lived in Europe for 400,000 years, eventually conquering fire, but became extinct in the years after *Homo sapiens* arrived from Africa.

The final step into biologically modern humanity took place around 200,000 years ago back in the East African Rift Valley where the origins of the family tree lay. A new species, *Homo sapiens*, emerged with a considerably larger front part of the brain. Physiologically this transformed the shape of the forehead from back-sloping to upright. Neurologically it expanded the parts of the brain responsible for judgement, planning, communication, memory, problem solving and sexual behaviour. Adam and Eve had arrived and would soon set about changing the planet forever.

Firelight was in many ways the midwife of our species, and remained our constant companion for thousands of years.

Hearths and Homes

I've built up the sequence of events described above by piecing together research by archaeologists, primatologists, evolutionary psychologists, paleoanthropologists, neuroanthropologists, and geologists, in particular the work of Robin Dunbar and Richard Wrangham. There are a lot of -ologists working on deciphering our deep history.[26] There are deep disagreements between the various scholars about several aspects of the story, and by necessity I've chosen between differing views at times. The overall story, however, is consistent with the current state of research.

A main area of scholarly disagreement is when our relationship with fire began. Many argue for a much later human relationship with fire than I've presented here, and it is true that there is no archaeological proof that hominids controlled fire before the earliest physical traces of human fireplaces.

Currently the archaeological evidence for hearths goes back 400,000 years in Europe,[27] 700,000 years in Israel, and up to 1.5 million years in Africa.[28] All of these dates are later than the emergence of *Homo erectus* 1.8 mya, which would mean *erectus* was the first species to use fire rather than *habilis*. However, methods for analysing archaeological artefacts continue to improve, and tend to push back the accepted age of the world's oldest hearths, so the archaeological record may change in the future. And bearing in mind the earliest fire sites would have been fairly haphazard, conclusive evidence of their presence may never be found.

Proponents of the earlier hominin relationship with fire, led by Harvard's Richard Wrangham, say that even without archaeological evidence, the biological evidence is clear. They maintain that various adaptations that transformed the species from *habilis* to *erectus* – the smaller gut, jaw and teeth, the larger brain, the longer legs and reshaped shoulders – in themselves provide compelling evidence of fire use.[29] That is, the lifestyle changes that resulted in *Homo erectus* – the implied move from climbing to walking and the change from a raw to a cooked diet – could *only* have come about through regular controlled use of fire.

In search of evidence for the link between firelight and socialising, a group of leading anthropologists led by Oxford professor Robin Dunbar came up with an ingenious experiment. They calculated how many hours in a day the various hominin species would require to do what they needed to survive, taking into account their physiology (specifically brain size and digestive system) and their habitats (ecology and group size). They reasoned that each individual would need at least a certain amount of time in a single day to travel, find food, chew and digest it, and rest, as well as bond with others and sleep. Using this 'time budgeting' approach the researchers calculated that *Homo erectus* needed sixteen waking hours per day to do everything it needed. With only twelve hours of daylight in their equatorial habitat, they must have found a way to extend the day into the evening, and that solution must have been firelight.[30] Practical, productive activities could take place during daylight hours, while equally important but slightly less light-dependent socialising could be moved into the evening, enabled by a campfire.

Whatever the precise timing, it is not contested that control of fire gave humanity illumination and protection, as well as warmth and cooking. It may also have given us our large brains, cooperative tendencies, language and social evenings: the foundations upon which we came to dominate the world.

As time went on, *sapiens* found more and more uses for fire: to bake clay into pots for storing and cooking food; to clear land to plant crops; to make bricks to build houses and temples and cities; to smelt metals into knives and swords and coins; to make steam to move pistons and power engines; and to drive generators to make electricity. Many of the major technological breakthroughs in history before the twentieth century depended on fire. And throughout all that time, firelight continued to flicker through the night, allowing us to cheat nature and see in the dark.

Modern humanity remains distinctly attached to fire. It continues to power much of our everyday technology, including its successor as our primary light source, electricity. While in most of its everyday applications fire is now hidden in engines and distant factories, we continue to associate a naked flame with our most treasured events. Candlelight and open fires conjure images of romance, friendship, comfort and intimacy. A campfire remains a much-loved institution, connoting images of freedom, childhood adventure and storytelling. We also continue to love eating flame-grilled meat, whether from a backyard barbeque or an international burger chain.

While few of us have a nightly campfire, we continue to gather regularly around a flickering light after dark, absorbing tales of bravery, tragedy, comedy and romance. Illuminated screens – cinema, television, and more recently digital devices – may be the modern equivalent of the campfire. Like our ancestors, we bond ourselves to our tribe and cement our identity through the stories we share around a flickering light – be they in the form of a song or a film or a video game – mostly after dark, and usually with others, real or virtual.

Fire began with life on Earth and is inextricably linked with cycles of life and death. Plants absorb the light and heat of the Sun in order to grow. Fire releases that light and heat from their corpses. Most of what we burn today is the fossilised remains of forests that covered the Earth long before dinosaurs existed. Their heat and light has been stored underground for hundreds of millions of years, only to emerge as oil, gas and coal, and be pressed into service by an upstart species whose brainpower, enabled by earlier fires, has turned it into an almighty beast with powers greater than those attributed to most of history's deities.

Fire was the first technology humankind harnessed in order to see more, by illuminating the dark. In controlling fire, the ancestor species adapted the environment to their will, rather than adapting themselves to the environment. In doing so, they became for the first time a creature apart from the other animals.

They became the People.

4

FROM THE EYE
TO THE PENCIL:
ART

Ah! would that we could at once paint with the eyes!
In the long way, from the eye, through the arm
to the pencil, how much is lost!

Gotthold Ephraim Lessing (1729–81), German philosopher,
cited in C.N. Douglas (comp.), *Forty Thousand Quotations:
Prose and Poetical*, 1917

Southern France, 1994

Jean-Marie Chauvet, Elliette Brunel and Christian Hillaire
spend their spare time roaming around the canyons of France's
Ardeche region searching for caves.[1] They are spelunkers –
cavers who love nothing more than squeezing through a tiny
undiscovered gap into an unknown space beyond. The trio has
discovered several important caves in the area, and each of them
is actively involved in the protection and preservation of the
many local cave systems. One day in the spring of 1994 they

were exploring the Estre canyon with three other caver friends, when they came upon something sure to quicken the pulse of any speleologist.

The Estre canyon is an abandoned meander of the Ardeche River in south-central France, next to the famous Pont d'Arc, a ten-storey-high natural stone bridge popular with canoeists, canyoners and other wilderness enthusiasts. It is dotted with the tantalising small caves, crevices, shelters and ledges that speleologists know might lead to a concealed entrance, or a secret passageway into an untouched realm, and the group had explored it many times before.

On this occasion the spelunking trio took their friends to see some prehistoric rock art they had discovered previously within the canyon. While heading back they stopped in a little rocky shelter familiar to many local cavers and performed the cavers' habitual inspection of all its nooks and crannies, looking for any little draft of air that might indicate a concealed opening, leading to a cave.

And sure enough, here in the shelter was just such a draft. It was just a slight current emanating from the rock face, but enough to get the group excited. They dug around the spot, moving rocks and earth looking for the source of the draft, but couldn't find it. After a couple of hours they concluded it was fruitless, just another *passe traou* from one part of the rock face to another. Glum and frustrated, the group left the shelter and went off to find more promising avenues of exploration.

Jean-Marie Chauvet left the site with his friends that day, but for some reason this particular draft – written off by the others as one false lead among many in the cavers' world – stayed with him. He tried to persuade the other two to go back to the shelter with him on several occasions over the subsequent months. Eventually, months later on a sunny December day a week before Christmas, he threatened to go back on his own if his friends wouldn't come with him. They relented, and the trio set off on a treasure hunt, looking more like coal miners than treasure hunters in thick overalls, heavy boots, and helmets fitted with headlamps. They packed the tools of their trade into backpacks: pick, shovel, rope, drill, mallet, punch.

From the deserted Pont d'Arc car park they set off up the ancient mule track that cut a diagonal across the face of the cliff above. After a while they veered off the track and followed familiar landmarks to their destination. The little shelter looked exactly as inauspicious as it had half a year earlier, but the little draft was still there, confirmed by the smoke of a lit mosquito coil.

The team started digging. For several hours they took turns eking out a passage through the rock. There was only room in the tunnel for one at a time, and it was a miserable job. The digger lay face down with arms stretched forward in a space barely big enough to wriggle into, hacking away at the packed earth in front of them with a mallet and punch. When they'd had enough, the other two would drag him or her out by the feet, cradling the broken rock and earth they'd managed to loosen in their folded arms, and the next person wriggled in to take their turn. They burrowed for about 20ft until the tiny hole became a narrow, winding tunnel. Finally the space opened out and Brunel, the smallest, squeezed through into a wider, higher passage where she could stand up. A few steps in she reached a sheer drop. By the light of her headlamp she could see nothing in the vast space ahead.

This was exactly what they hoped for.

After more hammering, squeezing, pushing and pulling, the two men joined Brunel in the passage. They gazed into the abyss before them, but the light of their lamps gave way to blackness after a few metres. What they needed was a ladder, and better torches. So they all squeezed back through the tunnel, walked down the track to the car, loaded up with kit, hauled it up the track, and squeezed themselves and the equipment back into the cave. They climbed down the ladder one by one and stood close together on the sandy floor of what appeared to be an enormous room.

The ground was level underfoot, and they inhaled the familiar, primordial smell of damp rock. Deep, dark silence engulfed them. Their torches revealed a magnificent cavern: a 50ft-high cathedral of twinkling white. Formations of sparkly, crystallised calcite leached from the limestone cliffs everywhere. Stalactites dangled from the roof, some as thin and delicate as strands of spaghetti, others pointed and menacing like daggers. Stalagmites grew up from the floor, thick and tiered like half-burned church candles, or touristy replicas of the Tower of Pisa. Some met in the middle to form whole columns, dividing the space into chapel-like chambers. Elegant wavy formations known as drapery hung from part of the roof. One cluster of three large stalactites suggested a trio of headless Motown singers in sequinned dresses. In another chamber empty nightgowns appeared to sway overhead in the moving torchlight. They saw that the cave continued, and set off to explore, leaving their boots behind to avoid damaging any formations on the ground.

As they moved through the cave, they saw the remains of animal bones and teeth scattered around. They recognised shallow pits dug into the sand as the nests of cave bears – a species that became extinct thousands of years ago. Suddenly, out of the silence, Brunel cried out, 'They were here.'

As her two friends turned towards her, their lamps illuminated the cause of her cry. Two red markings on the wall, man-made without doubt. Three stomachs turned over. They cast their lamps around and caught another glimpse of red on a dangling rock they had just walked straight past. A closer look revealed a small outline sketch of a mammoth, with a trunk and distinctive rounded skull. Then suddenly they caught sight of another image, a large cave bear in profile with ears, snout, chin and curved neck drawn in a single red line. There was more: a rhinoceros, a lion, more rhinos, more lions … all species long extinct here, but once native to the region. They saw imprints of painted hands and outlines of hands with paint sprayed around them. These were a common motif in cave art found elsewhere but they had never been seen in this part of France before. Likewise, the images of a hyena and a panther, and a bison composed of red dots.

The friends made their way past image after image, all fully aware that this was no routine discovery. They knew the paintings were extraordinary in content, quantity and preservation, and well over 10,000 years old. Chauvet, Brunel and Hillaire said later they felt like intruders, disturbing the spirits of the artists whose hands made these images, and remained, imprinted on the wall.[2]

When their lamp batteries started to fade, they returned home for a rest and some better lights, but they couldn't stay away long. They were back in the cave again later that night. This time they saw engravings: more animal shapes etched into the yellow wall by a finger or a tool, revealing the white surface underneath. They saw horses, an owl, another mammoth. The images had a graphic simplicity enlivened by simple details. They could be sketches by Picasso or Matisse.

They came upon a huge panel drawn in black charcoal taking full advantage of the curves and crevices of the rock face. The panel contained horses, rhinos, aurochs (ancient cattle), reindeer, and a bison with eight legs that looked like it was running. Most of the animals were in profile, facing the same way; many overlapped, giving the impression of urgency and movement, perhaps a stampede. There were two whiskered lions, looking cross with knitted eyebrows and flattened-back ears, in the middle of the scene. Did they cause the panic?

Another panel of charcoal drawings showed a group of sixteen lions, mostly just heads and forequarters, in what appeared to be a chase. Again they were in profile, all facing the same way with heads strained forward and eyes focused intently on something ahead of them. In front of them was a jumble of bison, mammoths and rhinos. One rhino seemed to have multiple versions of a horn, suggesting a violent movement of its head, or perhaps a group of several beasts in a row.

Their graphic style made many of the drawings look surprisingly familiar to modern eyes. Each species, for example, was drawn to a similar template, but individuals had different features, postures or expressions. Every rhino had cartoonish 'm'-shaped ears like a simplistic flying bird. Bear ears were curly, front-facing, three-quarter circles. Lion faces had a distinctive curved cheekbone and a white space around the eye. The animals' facial expressions were as anthropomorphic as those in a children's storybook. One bear actually resembles Winnie the Pooh, while the effect of the large dynamic panels is strikingly similar to the artwork in Disney's *Lion King* animation.

Some drawings had a more classical feel. A horse's head had had its charcoal outline enhanced by engraving. Several animals were shaded with charcoal to provide volume. Others had shading outside, lifting the animal off the wall in the manner of an Old Master sketch.

A century after the first European cave art was discovered on a farm in Altimira, Spain, Jean-Marie Chauvet and his two colleagues had found one of the most important examples of Paleolithic cave paintings in the world, and they knew it. They kept the secret over Christmas, then reported it to the French authorities. The world's leading cave art expert, Dr Jean Clottes, came straight away, and verified the cave's authenticity and priceless value in both archaeological and artistic terms. It was closed off to general visitors immediately and put under guard, as it remains today. The cave became known by the name of the spelunker whose persistence led to its discovery: Chauvet. In the spring of 2014 it was made a World Heritage Site, and a replica cave opened nearby in 2015.

What the three cavers didn't know at the time, and no one knew until radio-carbon dating was completed a few years later, was that the Chauvet wall paintings are almost twice as old as the famous cave paintings at Lascaux and Altamira. They were made by some of the first modern humans to arrive in Europe more than 30,000 years ago. These beautiful pictures were some of the earliest known examples of humans making art.

Early Stone Age

At least 2 million years before anyone entered the Chauvet cave, hominins had begun to modify materials to make simple tools out of sticks, bone and rocks. By the time *Homo sapiens* appeared 200,000 years ago these simple tools had become much more complex. Spears with stone tips and sharp stone blades were made using a multi-step process. Tools were made to be durable, kept and maintained, and carried long distances, in contrast to the earlier practice of making and discarding them as required.

Humans are not the only animals to use tools. Several ape species use stones for hammering, and dip stripped sticks into ant nests. Elephants pick up branches with their trunks and use them to scratch unreachable places, and have been known to plug waterholes with chewed up bark.

However, only humans use tools to make other tools. As far as we know, humans are the only animal to see something beyond what is actually there and have the imagination to think 'if I do that to this thing, it will be more useful'. In doing this, they are seeing not just with their eyes but also with their mind's eye.

Like harnessing firelight, tool-making represented a great leap forward in seeing the world. Using tools to make tools was evidence of abstract thought, and another example of seeing beyond how to adapt to the environment and instead adapting it to meet a need.

Critical as it was to the story of human development, in terms of the history of seeing, making tools was still fundamentally a utilitarian use of vision, as they made an obvious contribution to basic survival. What the spelunking trio discovered in the Chauvet cave was unnecessary to preserve life, and yet was entirely of life. The images they found were evidence of human imagination, and the first stirring of a vision-led culture.

We saw in a previous chapter that the first humans were different from previous hominins, especially in the shape of their heads and their brains, with a large neocortex that gives humans the ability to plan, make decisions and deal with complex social interactions.[3]

Homo sapien brains have another unique feature that might explain our propensity and ability to make images. Human brains undergo a growth spurt soon after birth, known as bulging, unknown in any other species. Palaeo-neurologists suggest this unique feature could be caused by the areas of the brain responsible for the link between vision, movement and memory known as visuospatial integration. In other words, hand–eye coordination and the ability to create and maintain a mental 'sketchpad'.[4]

At the same time as modern humans appeared, so too did signs of people making use of coloured clay for body decoration, beads made of shells, teeth and bones, and marks etched on stones and shells.[5] No one knows why they did this; perhaps they began as offshoots of practical activities. At some point 'creative' activities seem to have acquired a symbolic purpose, as a form of identification within a group, or between groups, or as part of a ritual such as initiation or marriage. They may have signified some form of status, achievement or ownership. In modern hunter-gatherer groups, body decoration and carving are used for all these purposes. The interesting thing isn't so much what these activities stood for, but the idea that they actually stood for something. For the first time a species was creating and using abstract visual symbols. Sight became more than a utility, for both the seers and the seen.

Depictions of real objects such as those at Chauvet is a mental and technological development far beyond these early artistic endeavours. Consider what's required to make a picture of something, even a picture far less complex and beautiful than the Chauvet cave paintings. Whether drawing from imagination or memory, the maker needs to hold an image of the object in their head. They need some sort of intention: are they simply trying to identify an object, to show what the object looks like, or to convey something more complex? To whom? Then they need tools: a mark-making instrument, a surface, and possibly some sort of pigment or colour. With tools in hand, the maker needs the manual dexterity to hold the instrument and, finally, the skill – probably only acquired by practice – to recreate the seen, imagined or remembered image.

Now consider the specific example of the Chauvet cave art. The images were made on rough surfaces with utterly primitive materials and by dim, flickering torches or grease lamps. The artists probably never saw the larger panels in their entirety. They could only have been working from memory, yet the images represent many different species at a high level of detail.

The people who made these images lived so far in time from our own that it is difficult even to conceive of it. We describe 'history' as the period since the first systems of writing developed about 5,000 years ago. The Chauvet cave paintings were made six *times* that long ago. There is certainly individual genius here, but it is inconceivable that these pictures sprang fully formed into the hands and mind of a single artist. There must surely have been some sort of artistic tradition on which these artists drew – a culture of painting and drawing where techniques were practised, developed and passed on. But there is no trace of this outside the cave, and no evidence of practice or preparation for the main drawings elsewhere in the cave.

It seems from the surety of the work here that whoever made these pictures knew exactly what they were doing. To have achieved the level of skill displayed, the individual artists must have spent hours, and probably years, observing animals and experimenting with mark making, using different materials, trying different shapes and shading. Art is also craft, and it will be an extremely rare modern-day artist who hasn't had years of training. Disney animation artists are still required to submit portfolios before even being considered as an intern or apprentice, despite the enormous role of computers in modern design work. The Winnie the Pooh illustrator, E.H. Shepherd, won a scholarship to the Royal Academy Schools in 1899. Picasso was formally trained in art from the age of 7. And 500 years before that, Leonardo da Vinci was apprenticed from the age of 14 to a famous contemporary painter and scholar.

Did the cave artists have master and pupil relationships? Were the most artistically talented singled out for a form of apprenticeship? Did the artists have an elevated status in the group, or some sort of shamanic role? We just don't know, as the only evidence we have that this type of art was even practised is what's on the wall, in finished form, deep inside the caves.

Which raises another question. While it is very lucky for us that these paintings were made deep in caves where they could remain undisturbed for thousands of years, what were they doing there? Why were these pictures so far inside the cave? Chauvet isn't the only example; many European caves are unmarked in the parts that enjoy natural light. The images tend to be further inside, beyond the light, where they could only have been made – and seen – by firelight. Just venturing that far into a cave was a brave endeavour, as it is today even with modern equipment. Quite apart from the dangers of getting lost, unexpected drops, rockslides and failed torches, cave bears still existed at the time of the paintings and, as the Chauvet spelunkers saw, had used the caves extensively.

Were painting and drawing ubiquitous? Did cave paintings survive only because they were hidden away, or are the caves themselves significant? Did the darkness, the silence, the isolation of the caves call forth a particular form of inspiration? What did the artists see in the caves?

Many theories on the meaning of Palaeolithic cave art have been debated during the century or so that it has been studied. One theory is that the pictures are simply representations made for pleasure: art for art's sake. Although there certainly seems to be an element of enjoyment taken in the Chauvet images, pleasure seems unlikely to offer a full explanation given their inhospitable location. On the other hand, perhaps the hostility of the caves is part of the point: one scholar believes much of cave art was the work of young adolescent boys, demonstrating bravado as they ventured into the deep darkness, and showing off their picture-making skills.[6] This may be the case with some cave marking but seems unlikely in the case of the accomplished works such as Chauvet's. Another theory is that the paintings were intended to invoke what is called Hunting Magic. When the artist depicts a desired outcome, such as a successful hunt, Hunting Magic makes the outcome a reality. Again, while this may be the case for some cave art, Chauvet cave has no depictions of animals as quarry, and none at all of humans (aside from some images that may be female pubic triangles). More recent theorists tend to reject the Hunting Magic thesis, while accepting that cave art probably does have supernatural purposes.

The theory most associated with the Chauvet cave is that advanced by Dr Jean Clottes, the expert who first verified the Chauvet art and who was the cave's chief scientist for many years afterwards. Clottes believes that prehistoric cave art was part of a shamanic religion, in which the cave walls represented the link between this world and the spirit world. Shamans sent their souls into the other world and interceded with the spirits on behalf of the human tribe. To this day shamans in small-scale communities are known to put themselves into trance states with drugs, chanting, fasting or sleep deprivation. Clottes and others believe that cave art may have been created while in these states, perhaps induced by fire, when the shamans 'saw' animals emerging from the spirit world on the other side of the cave walls.[7] This could explain why the paintings are so deep in the cave.

It is also possible that cave painters believed in animal spirits, without necessarily relying on a shaman to reach them. Yet another theory suggests the paintings were a means of apologising to the spirits of slaughtered animals, giving them a place to rest far from the everyday activities of the human herd.

Whatever the significance of the paintings to their makers, for us their mere existence, as the earliest examples of depiction, is surely significance enough.

Lookalikes

The Chauvet paintings display extraordinary mastery. They also show certain remarkable similarities to other cave paintings found elsewhere in Europe and in Africa, Asia and Australia. The oldest are in Sulawesi, Indonesia, and are estimated to be at least 40,000 years old. At all these sites animals are depicted similarly, despite being different species and the artists using different materials and techniques. They are typically in profile, with an emphasis on outline and especially the spinal contour. Heads are generally well defined, but legs and feet are not. An art historian might describe these works as belonging to a particular 'school', in which artists learn to take similar approaches to image making, but of course that idea is out of the question given the range of time and space involved. What else explains the similarity of these images?

Recent research suggests that this consistent artistic approach is down to the way the brain decodes visual information – that is, to the way we see.[8]

Prehistoric people lived among animals as both predators and prey, so identifying different species quickly was a critical survival skill. Any hunter, watcher or zoologist will tell you the side profile view is the best angle for animal identification as it gives the most basic identifying visual cues: spinal contour, head, eyes and neck. As we've seen, the first stage of recognising things visually often takes place unconsciously, out of the corner of the eye in vision's periphery. Fragmentary visual cues are processed and direct the viewer's gaze toward the object to make a full, conscious identification. Researchers believe that when depicting an animal, primitive artists drew on their most deeply embedded memory of each species – the outline profile view. This results in a very similar-looking image style across time, geography and cultures. Details such as filling and shading are not vital for identification and researchers believe these are culturally derived and hence not consistent across cultures in the same way as the outlines.[9]

Depiction takes vision beyond the here and now. It detaches the image of something from the thing itself, and from the experience of seeing it. In pictures, people can see things from far away or long ago. They can see things they have never seen in life, and things that don't exist. Infinite new visual possibilities open up. With depiction, the universe becomes more than the surrounding natural environment; it now includes the imagination made concrete. Depiction links the physical and the imaginary worlds.

Through depiction, memories and imagination can be expressed and preserved outside the original thinker's mind, and stimulated in the minds of others, even if they are miles away. Unlike sounds, such as spoken words and music, pictures once made can travel across time and space independent of their maker, just as the Chauvet images have travelled from the Ice Age to the twenty-first century. The same could not be done with sound until Thomas Edison invented the phonograph in 1877.

Oral songs and stories require an unbroken chain, person to person, to survive. Pictures can endure alone. They are the original time machine.

Europe, Last Ice Age

By the time the Chauvet artists were painting their caves, *Homo sapiens* had been around for tens of thousands of years. They had migrated north and east out of Africa as their numbers grew and arrived in Europe just a few thousand years earlier. The climate at that time was cold: this was the last Ice Age and there were thick glaciers covering much of the continent. People kept warm with heavy clothes and boots made from animal skins, and sheltered in shallow caves or lean-tos constructed from trees or mammoth bones covered in hides. Firelight from torches or lamps made from stone or clay showed the way after dark, as well as when people ventured deep into caves, as evidenced by the black marks where torches swiped the Chauvet cave walls and the hearths found on the cave's sandy floor.

As hunter-gatherers, Ice Age people would have spent their days outside in sunshine, all the more dazzling for being reflected off snow and ice. All their senses would have been called into constant use. Rain was rare as so much water was trapped in the vast glaciers, but people must have been acutely aware of other changes in weather conditions. They could surely sense coming changes in the nature of the air, or perhaps they recognised changes in cloud patterns foretelling weather shifts – a skill modern science didn't master for thousands of years.

The main quarry for Ice Age hunters was the large herbivores – horses, reindeer, aurochs, antelopes, bison and occasionally mammoths – that roamed throughout Europe in plentiful herds. Hunting was dangerous work, only made possible by lethal ivory- or flint-tipped spears, powered by spear-throwing tools that allowed the hunters to keep their distance from horns, tusks and hooves. Success at spear throwing required hunters to gauge the speed, distance and trajectory of the target animal's movement, all while being on the move. Once hit, a large beast could keep travelling for some time, taking the hunters further and further from their home encampment. When they finally killed the quarry, the hunters faced a new problem: getting the carcass back to the rest of the group without it being stolen. Kleptoparasitism, stealing another animal's food, is a common strategy among all carnivores, including humans. Carrying a massive carcass of fresh meat through a landscape populated by cave lions, cave bears, hyenas and leopards would not have been a good survival tactic. Perhaps some of the hunters would guard the carcass while others fetched the rest of the group, and they all made some sort of feast on the spot.

Or perhaps they had some help.

According to Penn State anthropologist Pat Shipman, early Europeans may have had a hunting partner.[10] Even though domesticated dogs did not appear in human settlements until 10,000 years ago, there is evidence from Ice Age sites up to 36,000 years old that humans and wolves may have hunted together and shared the spoils. Shipman's hypothesis is that semi-tamed wolf-dogs, with their superior sense of smell and faster running legs, tracked animals wounded by human spears and eventually surrounded or cornered them, and held them in place until the humans arrived with their spears to deal the final blow. This highly efficient hunting partnership kept the wolf-dogs safe from the tusks and horns of the wounded quarry, and saved the humans having to keep up with a wounded beast through miles of forest. The wolf-dogs would also be useful friends in guarding the carcass from potential thieves. Shipman contends that this finely tuned hunting machine was an unbeatable combination and outcompeted all other species, leading directly to the decline and eventual extinction of Neanderthal man and other large carnivores, including the cave bears, lions and hyenas that are so affectionately portrayed in the Chauvet cave.

Seeing is critical to this story. According to Shipman, the cooperation between Ice Age humans and wolves was only made possible because of the particular physical characteristics of the two species: their eyes.

We've seen that, uniquely among primates, human eyes are shaped to facilitate communication by gazing. Following a gaze is an important and instinctive form of human communication. Another species whose eye and facial structure enables its gaze to be followed is the wolf.

Like humans, wolves live and hunt in packs and have complex social structures. They also use silent gaze signals as a form of communication. Shipman's hypothesis is that at some point in the early years of *Homo sapien* residence in Europe, the two species found each other and developed a silent but lethal pairing. Using only their eyes, they managed to share their intentions, the risks of their endeavours and, ultimately, the rewards.

Perhaps this explains why wolves, like humans, are absent from the Chauvet cave paintings. Are they already seen as part of the human rather than animal sphere?

The theory of a human–wolf-dog hunting partnership is as yet unproven. Even if it is proved true, the practice may or may not have been common among those who dwelt near the Chauvet cave. On the other hand, inside the cave there are two pairs of footprints side by side, of a young child and a wolf. Some have speculated that the wolf was stalking the child, but there are no human bones in the cave. It is also possible the footprints were made at different times.

But I find it far more appealing to imagine a child, part of a healthy, strong family group, standing confidently next to a wild ally, neither pet nor subject but an equal partner in the vital enterprise of feeding the two packs. Why either of them would even be in the cave is yet another mystery, but having found themselves there I grant them a shared gaze, through flickering torchlight, into the eyes of a friend.

The Chauvet cave and its paintings let us see through the eyes of Palaeolithic artists and glimpse a fragment of their world: the shapes and forms of the animals that shared their living space. We can't know exactly what the artists intended but we can surmise that it was more than a simple representation. For both artist and viewer, sight was being called upon to interpret the world, both the world they could see and the world they could not. When the artists left the cave the animals on the wall would continue their dramas in the darkness, unseen and unseeable, as they would in a spirit world. Surely the paintings were made in pursuit of a quest to understand, communicate with and possibly influence the world beyond the one we see: actual spirits embodied in the animals perhaps, or the invisible powers of the universe they may represent.

There is a bitter irony in the Chauvet paintings. In the expression of the desire to capture the beauty, strength and power of the beasts on the wall, we see the seeds of man's eventual destruction of the creatures they so admired. Abstract thought, memory, creativity, ingenuity and dexterity – the keys to later human success – are demonstrated in abundance here. Cave lions and bears were outcompeted for game by humans and driven to extinction. The great aurochs, bison and horses survived at the expense of docile servitude to humans as meat and muscle. And wolves, perhaps celebrated as equals in their absence from the Chauvet paintings, eventually became dogs, the friendship of equals reduced to one of dependence and obedience.

Chauvet tells us that, more than a thousand generations before written history began, human sight was already much more than a survival tool. Imagination, supernatural beliefs, a desire for meaning and the exploration of the unconscious were up and running. Our mind's eyes had opened, and ascendancy of the visual had begun.

5

FROM EYE TO I:
MIRRORS

And how do I know who I am, until I see myself as
others see me?

Edmund Snow Carpenter (1922–2011), anthropologist, *Principles of
Visual Anthropology*, 1975[1]

In the famous story of Narcissus, a beautiful but proud boy falls
in love with himself when he sees his face reflected in a pool.
He can't bear to leave the image and pines away to his death
by the water, leaving only a little flower behind.[2] The story has
inspired artists for centuries, and his name has entered everyday
language as a term for anyone we consider vain or overly fond
of themselves, as well as describing a serious mental disorder.
The tale of the arrogant, enchanted youth captivated by his
own reflection speaks to two human fascinations: reflection and
seeing ourselves reflected. For thousands of years reflection was
the only route to seeing the face of the most important person
in our lives. As such, mirrors have played a profound role in
shaping the human psyche. Their enigmatic powers mean that

mirrors have also been critical to science, art, magic and the shadowy realms of the supernatural.

In the modern world, mirrors are so ubiquitous they can go unnoticed. Look again.

Turkey, 1961

The Konya plain in central Turkey rises 1,000m above sea level and is criss-crossed with fields of wheat and barley. The landscape is punctuated by the twin-coned peak of extinct volcano Hasan Dag, but it was a smaller double-headed mound that obsessed James 'Jimmy' Mellaart for nearly a decade. In 1952, Mellaart, a scholar at the British Institute of Archaeology in Ankara, was undertaking a survey of Anatolia and spotted the great mound in the distance. Lack of transport and a bout of dysentery prevented him from getting to the place known locally as Çatalhöyük – meaning forked mound – on that occasion. When he finally got back there in 1958, what he eventually found astonished him: 'Unmistakeable traces of mud-brick buildings, burned red in a conflagration contrasting with patches of grey ash, broken bones, potsherds and obsidian tools and weapons …' Most remarkably, 'these were found not only at the bottom of the mound, but they continued right up to the top'.[3]

Mellaart had discovered a huge Neolithic settlement, one of the first known towns in human history. He assembled a team and in 1961 began excavating. Over four seasons they uncovered nearly 200 houses, and thousands of artefacts.

Mellaart's dig revealed that Çatalhöyük was occupied for nearly 1,500 years, from around 7100 to 5600 BCE, by up to 8,000 people. The mud-brick houses they lived in were so close together that people got around by walking on the roofs and accessed the houses by climbing down ladders. Internal walls were plastered white and often decorated with hunting scenes or animal trophies – bulls and leopards were common themes. People used woven baskets to store food and pottery for cooking and made little figurines of human and animal figures from clay and stone. They kept their homes clean and replastered the floors and walls frequently, though they buried their dead under platforms inside their houses, and sometimes dug them up later and moved parts of them elsewhere. Periodically they carefully demolished a house and built a new one on top of the ruin: Mellaart discovered at least fourteen distinct layers of occupation, and later excavations revealed another four.

Çatalhöyük provides an extraordinary insight into one of the first settled communities, but there remains much about the lives of the people who lived there that is a mystery. One of the most intriguing artefacts Mellaart discovered was a small number of highly polished obsidian mirrors from about *c.* 6000 BCE, at the later end of Çatalhöyük's occupation. They give a good reflection in sunlight, not unlike what you see if you look into a smartphone with its screen turned off. They are the first known mirrors in human history.

James Mellaart was expelled from Turkey after a scandal involving suspected stolen artefacts, and the site was abandoned for nearly thirty years until a pupil of Mellaart's, Ian Hodder, obtained a permit for a new research project in the early 1990s. His international team started work on the site in 1993 and spent twenty-five seasons on further excavations.

Hodder's painstaking analysis of Çatalhöyük's remains has provided new insights into life in the proto-city. Despite the transition to settled life, the people of Çatalhöyük lived together in an interdependent and 'fiercely egalitarian' community.[4] Houses were of similar size and form, and there were no apparent public buildings or temples, nor signs of a ruling elite. Food and resources were shared – men and women ate the same things and did similar tasks.[5] There was no evidence of property ownership or differentiation of status. This equality pertained for about 1,000 years.

Then at a certain point, Hodder says, the Çatalhöyük community began to change. Households became independent and self-sufficient, specialising in certain activities and owning their own sheep and cattle, and trading with one another. They developed what Hodder calls a 'greater sense of individual self'.[6]

These subtle but fundamental societal changes took place at the around same time as the first appearance in Çatalhöyük of handheld mirrors.

Papua New Guinea, 1969

The large island of Papua New Guinea, located between the tip of Australia's Cape York and the Equator, is notoriously difficult to navigate. A massive chain of volcanic mountains runs its length from west to east, while the vast lowlands north and south of the mountains feature dense tropical rainforest and huge expanses of mangrove swamp. In the late 1960s the more remote mountains and valleys were still populated with isolated tribes living as people did thousands of years ago. Many had never had contact with Westerners.

In 1969 the American anthropologist Edmund Carpenter was a visiting professor at the University of Papua New Guinea in the capital, Port Moresby. Carpenter was an associate of Marshall McLuhan and had written extensively on modern media and visual culture. The Australian-administered government asked him to advise them on how they might use media to communicate with the indigenous population, including the isolated inland communities. They had been transmitting jolly radio news bulletins and messages promoting cleanliness and godliness to their geographically dispersed citizens, but these seemed to be having a negative effect. There had recently been protests and riots and the government wanted to understand where they were going wrong and put it right.

Carpenter viewed the assignment as a unique research opportunity. He wanted to explore how people reacted to seeing their own image for the very first time. He figured that the more remote tribes would never have seen cameras or mirrors and given the peculiar topography of their homeland may never even have seen their reflections in water.

Taking a small team, including a photographer and cinematographer who would later become his wife, Carpenter travelled to the high Papuan Plateau, home of an isolated tribe called the Biami who were believed to practise cannibalism. They looked the very model of 'primitives'. They wore huge quill rings in their ears, little wooden pegs through their noses and wooden bangles pushed to the upper parts of their sinewy arms. Their hair was cropped short around their faces and tied into extravagant arrangements high on the crowns of their heads. They wore woven loincloths and grass skirts with elaborate clasps at the groin.

Carpenter and the Biami approached one another cautiously and began communicating with gestures, sounds and facial expressions. After a while, Carpenter took out a large mirror and held it up in front of one of the Biami. The man was utterly dumbstruck. He put his hands over his mouth and ducked his head away in what looked like acute embarrassment. He moved out of the mirror's frame and looked at it from a safe distance. Then he came back and stood before the mirror, staring at his reflection. Carpenter said he was paralysed and transfixed, clearly troubled, his tension reflected in his twitching stomach muscles.[7]

When he put another Biami man in front of the mirror, his reaction was identical. Carpenter tried another, and another, and some of the women. They all responded in the same way.

Carpenter describes them as both exhilarated and mortified, in the grip of an intense personal dilemma:

> Once they understood that they could see their soul, their image, their identity outside of themselves, they were startled. Invariably, they would cover their mouth, and sometimes stamp their foot, and then turn away. And then [they would] take the image and look at it again, and hide, and so forth.[8]

Carpenter later described these intense reactions as 'the tribal terror of self-awareness'.[9]

Mirrors and the Self

Scholars and artists have long been fascinated by the relationship between mirrors and the idea of the 'self', both literally and figuratively. Sigmund Freud drew on the mirror as a metaphor and owned a collection of ancient Etruscan and Egyptian mirrors.[10] Carpenter's description of the Biami's reaction to the mirror seems uncannily similar to the 'Mirror stage' theory developed by French psychoanalyst Jacques Lacan in the 1930s and over subsequent decades. To describe it in very much simplified terms, Lacan proposed that when an infant first recognises him or herself in a mirror sometime in their second year, they go through a profound psychological change. The child develops for the first time a sense of themselves as an entity distinct from the world around them, creating their first mental image of an 'I', and the beginnings of the idea of a 'self'. This realisation is fraught with anxiety, however, because the child realises that they must separate from their mother, even while they remain utterly dependent on them. A version of this anxiety persists throughout life, the theory continues, because of the disconnect between the ideal 'I' of a person's imagination and the real 'I' reflected in the mirror.

In modern life, for better or for worse, reflections are everywhere. Mirrors are cheap and plentiful, and seeing our reflected image is, if not always a pleasure, almost as natural to us as waking up. The idea of being totally unfamiliar with our own reflection is almost inconceivable. But seeing ourselves face to face isn't at all natural. And if the reactions of the Biami are anything to go by, doing so for the first time can be extremely disturbing. What is it about the mirror image that prompts such intense feelings?

Imagine life in a mirrorless world. Everything you do and know is a function of how others respond to you. And everything you know about yourself is reflected in how you see others responding to you. 'You' are defined by how you perceive others responding to you. Your group or tribe is your only mirror. In such a world it is easy to imagine that group identity is supreme; group norms would be easily enforced and there would be little room for individuality that transgressed those norms.

A mirrorless society isn't necessarily indifferent to appearance, and certainly many traditional communities, including the Biami, spend considerable time and effort on grooming and adornment. Distinctive hairstyles, facial decoration, clothing and jewellery are important markers of identity. But, crucially, they signify group rather than individual identity. Without a mirror you can't groom yourself, and you can't see yourself groomed. Everyone has to groom each other. Grooming and adornment are mutual acts of kinship and bonding – trust in a satisfactory outcome is implicit – rather than expressions of personal identity or vanity.

Now let's introduce you to a mirror. When you see your reflection for the first time, it only takes a moment to recognise yourself. But you do more than simply recognise yourself. You suddenly realise how others *truly* see you. You've probably never really thought about it before. You see yourself from the outside and become for the first time 'self' conscious. You are an individual, a separate entity. So is everyone else.

What happens next? You start to think about that external, public persona, the 'you' that you now realise everyone else sees. You might change what you do or how you do it. Perhaps you start to compare yourself with others. Where do I fit in, you wonder? And look what else is happening: you're spending more of your internal, private time thinking about yourself. You're becoming more 'I', an isolated individual in addition to being a member of the group.

Now what if, just as you are going through this upheaval in your psyche, others around you are having a similar experience? Wouldn't relationships change as each person became more aware of themselves as a separate individual, and of each other as 'other'?

Did this happen to the Biami after they were exposed to a mirror for the first time? It is impossible to say for sure one way or the other, as Carpenter never returned to the Biami territory. Western culture forced its way into all of the remote Papua New Guinean villages over the following years, imposing a rapid and distorted 'evolution' on them. Today, the Biami people wear Western clothes, speak English and are embroiled in debates over traditional land rights, logging, road building and access to services such as education and healthcare.[11] Was there a line of cause and effect between the Biamis' exposure to the mirror and their subsequent adoption of Western-style individuality and 'self' awareness?

Carpenter tells of another tribe, the Sio, who were slightly less isolated when he met them in the 1970s. Carpenter took Polaroids and tape recordings of the Sio people and shared the images with them. Like the Biami's response to the mirror, they were initially astounded by the photos and sounds but soon got used to them. When Carpenter returned to their village *just a few months later* he found to his dismay that they had become unrecognisable. Men wore Western clothes, people walked and acted differently, and some had left the village. Carpenter felt his visit had tipped the scales, and in one brutal movement they had been torn out of tribal existence and transformed into detached individuals, lonely, frustrated, no longer at home.[12]

He never completed his project for the Port Moresby government, and questioned his actions for the rest of his life.

Near East, *c.* 3000 BCE

Did something like the Biami's 'mirror moment' happen to the population of Çatalhöyük thousands of years earlier? Did some artisan create the first mirrors and prompt a profound change in the way people saw the world? Or was change already under way and the invention of mirrors was one response to it? Was there any link between the appearance of mirrors at Çatalhöyük and the shift towards a more individualistic society? It is a tantalising possibility.

What we do know is that over the next 2,000 years the world's first true civilisations arose in Mesopotamia and Egypt, both within trading distance of Çatalhöyük. The idea of owned property was established, leading to both wealth and poverty. Inequality of wealth became inequality of status. Eventually distinct classes of citizen – ruler, aristocrat, priest, scribe, artisan, slave – emerged, each with an assigned status and role based on the accidents of their birth and particular capabilities. These societies were anything but egalitarian.

We also know that they used mirrors. They were made of copper and, later, bronze, and sometimes gold or silver. Metal mirrors have been found from around 4000 BCE in Mesopotamia and 3000 BCE in Egypt. From these times Mesopotamian and Egyptian paintings and sculpture included many depictions of nobles, often women, holding mirrors while being attended to by servants.

The appearance of metal mirrors also coincided with a marked change in how women were depicted. Neolithic female images typically portrayed large and round women with full breasts and hips, and often a prominent vulva. Facial features were obscure or absent altogether. These images, usually carved figurines, are often called Mother Goddesses, as they seem

Fig. 1 Detail from the sarcophagus of *K3wi.t*. Dynasty XI, c. 2050 B.C.

Fig. 2 Detail from the superstructure of tomb 24 at Sheik Said. Dynasty V, c. 2450-2300 B.C.

Fig. 3 Detail from the Tomb of Mereruka at Saqqara. Dynasty VI, c. 2300-2250 B.C.

Three depictions of mirrors obtained from Egyptian tombs. 2050 BCE, 2450 to 2300 BCE, and 2300 to 2250 BCE respectively.[13]

to celebrate robust fertility as the ideal of femininity. Similar images have been found in Europe, at Çatalhöyük, in Neolithic Mesopotamia and pre-dynastic Egypt.

As settlements turned into civilisations, female images changed. Bodies portrayed became younger and slimmer. Facial features were much more distinct and conformed to an ideal-ised norm any of us would recognise as beautiful: symmetrical features, large eyes and full lips often emphasised by cosmetics. The modern ideal of femininity, in which youth and aesthetic beauty are uppermost, emerged. And it was achieved with the use of a mirror.

What prompted this change? As always, there is probably no single reason. But here's a plausible scenario. As we've seen, pre-settled societies were (and are) typically 'fiercely egalitarian' with equal roles for men and women. Survival of the group was the first priority. A fertile woman literally embodies this objec-tive, so it makes sense that this was how women were celebrated and idealised.

As agricultural civilisations developed, the roles of men and women diverged. The physical demands of farming and pro-tecting settlements from war raids gave men – usually physically stronger than women – dominant roles in society. Property and status rose in importance, and the wealthiest started to acquire assets beyond their basic needs, including beautiful things. For the most dominant men, women became one such asset. A rich man had a beautiful wife, perhaps several, to provide him visual and sexual pleasure as well as children, and as a medium to display his wealth and status to other men. The notion of the 'bride price' – a payment from the groom to the bride's family in return for her hand in marriage – first appeared around this time. This made beauty a valuable currency for a woman and for her parents.

The moment when a woman is most 'valuable' in this sce-nario is no longer the moment she gives birth, but the moment of the transaction passing her from father to husband – her marriage. Thus the 'ideal' woman is one just coming to marriageable age, with a beautiful face and a virginal body unravaged by childbirth.

How did a woman maximise her own asset value? By maximising her beauty. And what did she need to do that? A mirror and cosmetics. Hence the association of women with mirrors, which were now a tool of self-interest as well as self-knowledge.

In the civilisations of Mesopotamia, the highest goddess was Inanna, also known as Ishtar. She represented fertility, love and war manifested in an image of youth, beauty and sexual desirability. The portrayal of women as sex objects has a long, long history.

In Egypt, the ultimate example of the idealised woman is one of the most famous images in the world. Nefertiti, the Egyptian queen and mother of Tutankhamun, was depicted as tall, slender, with a long graceful neck and perfectly shaped, enigmatic features. She could grace the cover of *Vogue* or a Milan catwalk, so exactly does she capture the modern idea of beauty. Archaeologists have been searching unsuccessfully for her tomb for many years. If they find it intact, I am sure it will contain plenty of mirrors.

Greece, *c.* 500 BCE

The link between mirrors and identity, selfhood and beauty continued into the Classical period. By around 500 BCE the concept of the self was not only explicit, it was embraced. The ancient Greeks inscribed the maxim 'Know Thyself' onto the Temple at Delphi, one of their most sacred buildings. They believed self-knowledge was the apex of all knowledge, and could be assisted by the judicious use of a mirror for self-contemplation. The philosopher Socrates, who lived around 400 BCE, considered mirrors an important tool in 'moral' education, and was said to have proclaimed that young people should look at themselves in a mirror frequently in order to 'adapt their behaviour to their beauty, if this were the case, or to hide their defects through education'.[14]

A few hundred years later, the Roman philosopher Seneca also wrote about mirrors at length. He both deplored and applauded their use. Echoing his Greek predecessor, Seneca declared that mirrors were invented so that man would 'know himself' and in so doing acquire wisdom. Understanding whether one was handsome or ugly, young or old, should guide one's behaviour, he said. An old man 'can see his white hair and refrain from activities unsuitable to his age', for example.[15]

Seneca acknowledged a risk that mirrors could be directed to vain or even sordid ends. He gave the example of Hostius Quadra, a rich and decadent aristocrat. Hostius enjoyed the carnal pleasures of both men and women, and Seneca described in graphic detail how he used strategically placed mirrors, including magnifying mirrors, to observe himself and his partners in the course of various sexual acts. He took particular pleasure, Seneca said, in the illusion of great size his mirrors presented.[16] In so doing Hostius abused the power of the mirror.

But there's the rub (no pun intended). It is an inevitable consequence of self-knowledge that those who possess it will come to recognise their own desires. Once a desire is recognised, the logical response is to satisfy it. The itch, once acknowledged, demands to be scratched. It is surely no coincidence that formal laws appeared at around the same time as a greater sense of self-awareness emerged. Whereas traditional societies regulated themselves with unspoken group norms, as people became more individualistic, formalities such as laws and punishments were required to suppress certain behaviours. Self-awareness has a dark side, which is why laws and sanctions exist, and also why they are consistently ignored.

Present Day

Hostius' self-regarding voyeurism is by no means unique – think of the cliché of a mirrored ceiling above the bed in a sleazy motel. But psychologists have found that seeing our image in a mirror can also have a regulating, rather than corrupting, effect on behaviour. In one 1970s study, college students were asked to complete a fake intelligence task. One half were seated in front of a mirror and had a recording of their own voice playing in the room during the 'test', to make them highly self-aware, while the other half didn't. Those who didn't see and hear themselves while doing the test cheated on the test ten times as often as those who were self-aware.[17] In another experiment, several hundred children out trick-or-treating on Halloween were greeted by a host, who told them to take only one piece of candy each, then left the room. In half of the cases, a mirror was placed on the table where the children could see themselves, and in the other half there was no mirror. It turned out that when the mirror was there the children were half as likely to cheat (take an extra candy) as when it wasn't.[18]

These experiments demonstrate a version of the 'looking-glass self' theory developed in the early twentieth century by sociologist Charles Cooley. He believed that our 'self' is made up of what we think others think of us. The theory goes that when we are made self-aware – perhaps literally, by seeing ourselves in a mirror – we are reminded of the person society expects us to be and are more likely to act accordingly. As

Cooley put it: 'I am not who you think I am; I am not who I think I am; I am who I think you think I am.'[19]

So mirrors can impose themselves on human psychology in many ways. In Cooley's world, mirrors act as a virtual tribe, imposing societal norms. In Carpenter's world they stimulate self-interested and individualistic behaviour. In Seneca's they unleash a wild and corrupt ego. If this feels contradictory, I agree. But who said scientists ever agree, and who said human nature was straightforward?

Mirrors are still being used to study the human psyche today and psychologists remain keenly interested in human interactions with their own reflection and what they say about the self and human consciousness. Children start to show an interest in their own mirror image from about the age of 6 months, waving, smiling, touching and otherwise interacting with it. From about 15 months to 2 years they begin to recognise their own image in the mirror and will put their hand on an unusual mark on their (real) nose when they see it in their reflection. This positive response to what has become known as the 'mirror mark test' is an important milestone in a child's development and, in a refinement of Lacan's theory, coincides with children developing traits related to self-awareness such as empathy and recognising their carer as being separate from themselves.[20]

The mirror mark test has been applied to other animals to investigate whether they possess self-awareness. Some chimpanzees and orangutans have passed, as has a single elephant. Seneca might be interested to learn of the reaction of two male dolphins, Pan and Delphi, who, when put through a mirror mark test, went into something of a sexual frenzy.[21] In a half-hour period in front of a mirror, Pan tried to have sex with Delphi twenty-four times and Delphi tried to enter Pan nineteen times. If the pair drifted out of the mirror's line of sight they stopped the sexual activity and repositioned themselves for a better view. Eat your heart out, Hostius.

Venice, *c.* 1300

Romans invented glass mirrors but they were prohibitively fragile and expensive to become common objects. Metal mirrors remained the main instrument of reflection for centuries. It wasn't until around 1300 that improvements in Venetian glass-making technology made clear glass mirrors available. They were still expensive, well beyond the reach of ordinary people, but affordable to aristocrats and the rising merchant classes. 'Looking glasses' spread around Europe over the next two centuries.

What do you know? The spread of glass mirrors coincided with a notable rise in individualism from around the fifteenth century.[22] Portraits became extremely popular. People commissioned paintings of themselves for their homes and in art commissioned for churches. 'Donor portraits' depicted those who gave large amounts of money to their church in painted religious scenes alongside saints, apostles and the Holy Family. Artists explored their own image in another new genre, the self-portrait. The trend began with Jan Van Eyck's 1433 image of a man with a red turban, and has been a recurring theme of artistic self-expression ever since. And because art imitates life imitates art, mirrors themselves featured in many portraits, perhaps most famously in Jan Van Eyck's *Arnolfini Portrait* of 1434, which features a convex mirror at the rear of the scene reflecting the backs of the painting's two main subjects, and a tiny glimpse of the artist himself. The painting's exquisite detail echoes the uncanny miniaturisation a convex mirror produces, and some scholars further contend that Van Eyck (and later artists) used concave mirrors as an optical drawing device (on which more is said in a later chapter).[23]

Catoptrics

Of course, mirrors don't just reflect people. They reflect light in all sorts of ways. The science-loving Greeks were the first to study the properties of reflection, known as catoptrics from the Greek word for mirror. Euclid described the geometry of plane (flat) and convex mirrors around 300 BCE, while the famous Greek mathematician and inventor Archimedes, born a quarter of a century after Euclid, is supposed to have used an array of mirrors to focus the sun's rays onto an approaching Roman fleet, creating a sort of 'death ray' that burned them into the sea. The popular US television programme *Mythbusters* recently tried twice to replicate this achievement without success,[24] concluding the Archimedes story was indeed a myth, busted. That didn't stop the writers of James Bond film *Die Another Day* introducing the 'Icarus', a satellite with an array of mirrors that could focus the sun's rays into a powerful light beam. Naturally, the villain had the Icarus beam chase Bond across an ice sheet, melting the ice in its path and shearing off the side of an ice cliff. A trio of physicists from Leicester University estimated that the power required for a real-life Icarus would be more than 500 times total world power, and that the effect shown in the film would require mirror panels with an area nearly *2.5 million times* that of the mirrors on the International Space Station.[25]

After the Classical period, the science of reflection – and much else – was forgotten for centuries until the late Middle Ages, when scientific texts from the Arabic world made their way back to Europe. In 1425 or thereabouts, the Florentine goldsmith, engineer, architect, painter, sculptor and all-round genius Fillippo Brunelleschi used a newly available flat glass mirror in an experiment that revolutionised art and, eventually, science. Standing at the cathedral door, he painted a picture of the Florentine Baptistry opposite onto a small board. He drilled a hole in the board at the picture's theoretical point of view and had a viewer stand in the same place and look at the Baptistry through the hole, from the unpainted side. Then he passed a flat mirror in front of the board so that it reflected the painted side. To the viewers' amazement, the painted image reflected in the mirror was virtually identical to the real Baptistry.

Brunelleschi had rediscovered the technique of geometric linear perspective that had been lost since Roman times. It was quickly copied around Europe and resulted in a new, realistic painting style that supplanted the typically flat, symbolic art of the Middle Ages. Leon Battista Alberti codified the principles of linear perspective in detail in his famous 1435 book *On Painting*, crediting Brunelleschi for 'discovering' it, and the method was adopted by architects and engineers. With the ability to make technical scale drawings they could thenceforth invent, improve, and correct the most complex buildings and machines without having to waste time and money building and testing their ideas with three-dimensional models.[26]

The rediscovery of linear perspective, made possible in part by the invention of flat glass mirrors, was a defining element of the Renaissance. The rediscovery of the scientific underpinnings of mirrors and reflection was another important shift in how people saw the world at the time: it was the beginning of a quest to understand the workings of the natural world in terms that went beyond spiritual or mystical explanations.

Reflection has advanced science in many other ways. Two and a half centuries after Brunelleschi's mirror experiment,

Newton (b. 1643) built the first reflecting telescope, and mirrors are still used to see deep into space today, including on the famous Hubble Space Telescope. Mirrors are also critical to microscopes, cameras, lasers and modern high-definition TVs, contributing to many later chapters in the history of seeing.

Magic Mirror …

Mirrors have always been associated with the world of magic, both in conjuring tricks and deeper occult activities. The reflected world of the mirror's image is a step removed from reality, the polished surface creating a veil between reality and its image. Are the two worlds identical? Looking in a mirror I see something move, behind me. I turn around to catch it … and it's gone. The mirror plays tricks on our perceptions – we see depth even when we know the mirror is flat. What else is it hiding? And as we've seen, mirrors weren't always as perfectly flat and clear as they are now. If one looks into an ancient mirror, the possibility of magic, or a link with worlds beyond our own, becomes quite easy to see, even for a twenty-first-century sceptic.

Look into a resting smartphone screen. If there is plenty of light, you'll see a pretty good reflection of your face. But if you tilt it slightly and look behind you, the mood changes. I challenge you not to find the image unsettling. If your room were illuminated by flickering candles or oil lamps, as the rooms of the ancients would have been, you might well imagine supernatural forces within the screen.

The imperfections of metal mirror images are also suggestive of worlds beyond, especially as they were often made slightly convex in order to give an image of the whole face. Try looking at yourself in the top of a saucepan lid. You will see a reasonable image of your face but your surroundings will look decidedly eerie.

Conjurers, magicians, charlatans, theatrical impresarios and special effects artists have long employed mirrors to entertain, deceive and amaze their patrons. We still laugh at our distorted image in a fairground mirror, lose ourselves in a mirror maze, and allow ourselves to be terrified by spectral figures appearing from nowhere in a haunted house. Nineteenth-century audiences were scared out of their wits by phantasmagoria shows created by projecting images onto smoke using mirrors. Twenty-first-century audiences were just as amazed when a similar technique conjured Michael Jackson onto the stage at the Billboard Awards five years after his death. Floating coins, levitating assistants, vanishing body parts – all illusions that can be created with mirrors.

Reflections and mirrors have made their way into folklore, fairy tales and everyday superstitions, and they often had religious significance. One widely held belief was that mirrors capture the soul. In some cultures, people cover the mirrors in a home where someone has died to prevent the soul of the deceased from going the wrong way on its path to the afterlife and getting 'stuck' in the house. Medieval pilgrims often took mirrors to pilgrimage sites to hold up to holy relics, believing the mirror could capture and retain the relic's holy powers. The idea of the mirror as reflector of the soul is also the reason why vampires and witches – soulless both – don't show up in them.

The idea of the soul in residence is also the origin of the superstition that you will have seven years' bad luck if you break a mirror: breaking the mirror disturbs the soul embodied in the reflection, and renewing it takes seven years. This seems harsh, but it is a minor punishment compared with the fate of a traditional Scot, who would face imminent death if they broke a mirror.[*]

---

[*] A more prosaic explanation for the seven-year superstition is that this is how long it would have taken a servant to pay back the cost of a looking glass if they were unfortunate enough to break one.

In the British Museum there is an Etruscan bronze hand mirror dating from 200–300 BCE with the word 'SUTHINA' roughly engraved on its reflective surface. The coarse lettering looks like lipstick scrawled on a hotel mirror in a B movie and is in marked contrast to the finely detailed illustrations on the mirror's reverse. It is strangely unsettling even in a room filled with relics of the long dead. *Suthina* means 'for the tomb'. It wasn't a curse levied on the mirror's owner; the inscription was intended to make the mirror unusable so that it wouldn't be stolen by grave robbers. Mirrors were included as 'grave goods', and buried with their owners, in many cultures. There are many theories on why this was so: to keep the deceased looking their best in the afterlife, or simply to allow the deceased to hang onto a cherished item. Or perhaps the mirror was there to keep the soul safely in the afterlife, and stop it wandering back into the world of the living.[27]

The fairground mystic gazing into her crystal ball, and Snow White's evil stepmother quizzing her magic mirror are both 'seers', practising the art of scrying. Looking into mirrors for the purpose of seeing the future, or the truth, has been going on for millennia. Ancient Egyptians gazed into bowls of dark ink; Mesopotamians used oil. In medieval Persia wizards looked into the Cup of Jamshid to see all the seven layers of the universe, while the famous French seer Nostradamus (b. 1503) looked into a bowl of water for his visions of the future. Queen Elizabeth I's close advisor, John Dee (b. 1527), with the help of other scryers, used a crystal to communicate with angels in his for search for truths about the workings of the natural world, having exhausted earthly sources.[28]

None of these people were seen as cranks at the time, nor were they operating outside their predominant religions. Nostradamus and Dee were both devout Catholics. In modern times, however, scrying tends to be associated with the dark arts and crackpots. Aleister Crowley was a famous occultist of the early twentieth century, and used reflections in a golden topaz to scry his way into other, forbidden, worlds. When someone called him 'the wickedest man in the world' a judge rejected Crowley's libel claim on the basis that the statement could not be proved untrue.

The Amazing Mirror

Commonplace as they are today, it doesn't take much to be bewitched by the mirror. It is a solid shapeshifter, both a miracle of illusion and a source of the truth. Loyal friend one minute, tyrant the next. Imparter of good news and bad, truth and lies, corruptor and invigilator both. It has been used to connect man with the gods and to endow him with god-like powers. It can distort and it can clarify, take us into the deep past or, perhaps, reveal the future. Unknowable, yet governed by strict laws, the mirror is a many-faceted paradox. It joined the human story thousands of years ago as a rare and precious item. Today, the mirror is cheap, ubiquitous and taken for granted. But it continues to exert its irresistible power over us.

Its very name provides a clue. The word 'mirror' comes from the Old French *mirer*, which in turn is from the Latin *mirari*, which means 'to look at in wonder or awe', or 'to be amazed'.[29] That seems an appropriate term for a device whose physics have been understood for thousands of years but can still surprise and baffle us. A mirror takes the three-dimensional world, with all its many colours, shapes and depths, and makes it flat. It looks behind us and shows us what we can't ordinarily see. But what is most fascinating of all to humans – that most fascinated of species – is that the mirror lets us see what nature and evolution kept from us, but which is inextricably entangled with how we see the world and everything in it: ourselves.

When Snow White's stepmother asked the magic mirror on the wall 'who's the fairest one of all?' the image reflected back told her all she needed to know without the need for a disembodied, Disneyfied voice. For many years she could see that she was the most beautiful woman in the land – the mirror told her so. It also told her when her ageing face was no longer lovelier than that of the newly matured Snow White.

When I look at my face in a mirror I sometimes glimpse my late mother looking back at me; at other times I see my daughter in thirty years. The mirror indeed transports me, to see both the past and a vision of the future. Place two mirrors facing one another and you see infinite versions of yourself; place them another way and you can make yourself disappear. Stare into a dark mirror long enough and you may find your mind wandering into your subconscious, where lurks who knows what. The secret powers of the mirror are the natural miracle of reflection combined with the human imagination.

Not magic, but magical.

6

GEOMETRY
OF THE SOUL:
WRITING

Writing is the geometry of the soul.

Attributed to Plato (*c.* 429–347 BCE)

Bagastana, Persia, Summer 1836

The *biganeh* – 'stranger' in Persian – stands facing the sheer rock wall, so close he can taste the heat radiating from the 100 million-year-old limestone. His jaw is clenched, cheeks flushed with the combination of heat and concentration. His legs are planted solidly apart, arms raised just above his head. Khaki shirtsleeves rolled to the elbows reveal strong forearms, tanned to the red-brown shade of a fair complexion subjected to strong sunlight. His hands are large and capable, smooth – a gentleman's hands marred only by the calluses of a horseman. His left hand presses a leather-bound notebook open against the rock face, index finger marking a place for the pencil he holds in his right. His head is wrapped in a *chafiye* – a light-coloured cloth like those worn by the locals; he looks up frequently to a fixed point above him, and back down to the notebook.

Slowly and meticulously he fills in lightly ruled lines with tiny wedge-shaped marks, copying the intricate inscription carved into the wall inches from his face. Some of the marks are tilted, some horizontal, some chevron shaped, some in combinations similar to an F or an H. They form a script he knows as cuneiform, but they mean nothing to him, making the job of copying them harder. With no familiar points of reference, he must be scrupulously accurate. Despite the physical awkwardness, the heat and the apparent monotony of the task, the young man holds this position for more than an hour.

Finally he relaxes his stance, rotates and stretches his arms and shoulders, and turns carefully away from the wall. He sits down where he is, on a ledge little more than a foot wide, 300ft above the ground. Dangling his bare feet over the side, he takes a draught from a calfskin water bag, balances a light pith helmet on top of his *chafiye*, and reaches into the pocket of his short trousers. Retrieving a small knife, he sharpens the pencil to a fine point.

The ledge he sits on cuts across the lower part of a vertiginous limestone rock face that continues up a further 1,500ft. On the elevation immediately behind him are five huge panels, each about 15ft high and 6ft across, all inscribed with cuneiform characters. Immediately above these panels is a huge relief sculpture, 10ft high and 18ft across, showing a king and two attendants addressing a line of prisoners roped together around the neck, and standing over another vanquished prisoner who appears to be begging for mercy. The scene is watched over by a winged figure, presumably a god. On either side of the sculpture are two further inscriptions, each in a different version of the wedge-shaped script, although this would not be apparent to the casual observer. At his present level there are three more inscribed panels to the right, in a third version of cuneiform text.

In all, the monument is 25ft high and 70ft wide, and is known as the Behistun Inscription. It is carved into the rock of Behistun, on the sacred mountain of Bagastana within the Zagros range in today's Iran. The massive limestone and dolomite Zagros range was created eons ago by the collision of the Persian and Arabian tectonic plates, and it has for millennia formed a natural border between empires. In 1836 the Persian Empire lies to the east of the mountains, the Ottoman Empire to the west.

The *biganeh* takes in the sweeping plains before him. At the foot of the mountain his local manservant, guide and groom wait patiently with several horses. They stand in a shaded spot next to a sacred spring that bubbles out from the base of the cliff into a welcoming pool. He calls out to them in Persian and they wave up at him, then say something among themselves that he can't hear, and laugh. They are probably telling one another, again, how crazy this Englishman is. No one climbs that cliff.

As he watches them, a black-clad woman arrives from the nearby village carrying a large clay jar on her head. She fills it at the spring, lifts it back onto her head, and turns to walk back to the village. He thinks briefly of the native girl who shares his bed in the house where he lives at Kermanshah, 30km to the west along the dusty road that crosses his view just beyond the spring. The Royal Road, as it is known, was once the link between the ancient Mesopotamian capitals of Babylon and Medea, and one of the Silk Roads between China and Europe. In 1836 it is still an important trade route, linking Tehran – capital of Persia – with Baghdad, an outpost of the Ottoman Empire.

To the right of his view, about half a mile away, is the Caravanserai: a large, low rectangular structure with a domed main building and a series of stables and simple rooms facing into a large central courtyard. He can't tell from here how many Persian, Arab or further distant traders and travellers are resting there today, as they have for centuries, surrounded by their heavily laden camels, donkeys and horses.

Beyond the spring, the road and the Caravanserai are the fertile plains of Kermanshah, a patchwork of haphazardly planted summer crops and colourful wild flowers. A narrow stream snakes out from the spring-fed pool below and winds its way across the flat plain stretching for miles before him until it is halted abruptly by another bare rock mountain in the distance.

After a short break, he stands, turns back to the wall and resumes the painstaking task of copying the intricate inscriptions that cover the surface before him. Of the 1,119 lines of cuneiform text carved into the Behistun rock, 26-year-old Englishman Henry Creswicke Rawlinson, lieutenant of the British East India Company Army, will today manage to copy only a few. He will return here again and again when his army duties allow over the next eleven years, each time risking life and limb to climb the rock and record a little more of the ancient text. He harbours a dread fear of obscurity, and believes cuneiform is his opportunity for fame and glory.

Henry Rawlinson's burning ambition is to be the first man to decipher the world's earliest writing system.

The Earliest Writing

Some 700km south-west of Behistun, midway between Baghdad and Basra in modern Iraq, lie the ruins of Uruk. Around 5,000 years ago it was the world's first true city. Uruk was situated on the banks of the Euphrates River in Sumer, towards the south of the broader region known as Mesopotamia, meaning 'land between two rivers'. This area of lush alluvial plains between the rivers Tigris and Euphrates curves north-west from the top of the Persian Gulf to the eastern Mediterranean. It is often called the cradle of civilisation.

Humans started growing cereals, keeping animals and creating settled communities in Anatolia, where Çatalhöyük is situated, around 10,000 years ago. Farming seems to have slowly spread east and south around the Mediterranean, and through Mesopotamia from north to south. Farmers started to keep track of their seeds and yields and, perhaps as long as 8,000 years ago, a system of counting developed using small clay tokens. The tokens were shaped differently to represent different items such as a basket of grain, a sheep or a jar of oil. Over time, tokens were embellished with marks so they could represent a wider range of goods.

There was a direct relationship between the tokens and the things they signified; numbers as an abstract concept did not yet exist. Clay tokens were used all over Mesopotamia for thousands of years and may still have been in use as recently as the first millennium BCE. Archaeologists have found tens of thousands of them in Palestine, Syria, Anatolia (Turkey) and Iran.[1]

Over the centuries, agricultural and social practices grew more complex and diverged, and different cultures developed. In Mesopotamian settlements a temple dedicated to a particular god and run by a priest or priestess was the centre of village life. Villagers would bring all their surplus grain, livestock and products such as oil and wine to the temple, where they would be stored to provide offerings to the deity or distributed later in times of hardship or shortage.[2] Priests used clay tokens to keep track of goods going into and out of the temple.

Eventually people in villages and towns started to specialise in different activities. Artisans made pottery and metal tools.

Some farmers focused on particular types of food production such as herding, grain growing, or fishing. As specialisations emerged, goods and services had to be exchanged and traded. As things became more complicated, priests and traders needed a way to track agreements, exchanges and receipts. In Uruk, around 3500 BCE, a system developed whereby a set of clay tokens representing a particular transaction would be wrapped together in a clay sphere about the size of a tennis ball called a bulla. The bulla provided a secure record of the transaction, but had one major shortcoming: the tokens inside couldn't be seen without breaking it open. The first solution to this problem was to press the tokens going into the bulla into the outside surface of the clay ball while it was still soft, making a visible impression of the ball's contents.

At some point, someone realised that storing the tokens in the bulla was not really necessary: they could keep a record of the transaction by simply making impressions of the appropriate tokens on a clay tablet. Having made that conceptual leap, another someone realised that the impressions on the tablet did not need to be of the tokens themselves, but just needed to represent the tokens in a recognisable form. Or rather, the impressions needed to represent the items that were signified by the tokens, thus taking the token out of the equation altogether. At this stage people started using a cut reed to make it easier to etch symbols on the tablets instead of impressing the tokens themselves.

This record keeping and accounting technique quickly developed into a system known as proto-cuneiform, the world's earliest form of writing and numerals. Proto-cuneiform used small clay tablets about the size of credit cards, divided into rectangles by lines. Within each rectangle a scribe would impress tiny symbols representing different goods, quantities, names, places and dates.

The following translation, from a tablet found in Uruk, gives an example:[3]

First side:
1 billy goat, 25th day
146 various sheep and goats
26th day
delivery
Urkununa
Reverse:
Accepted
Month 'big festival'
Year: Amar-Sueen destroyed Urbilum

The progression from token to tablet took 4,000 years, but once proto-cuneiform emerged around 3,500 BCE it advanced quickly as new signs were added to those derived from tokens. Many of the new signs were pictograms – standardised images of the object they represented. An ox's head represented an ox, for example. Other signs were more symbolic: the goddess Eanna, for example, was depicted by a gate and post.[4]

From the early days of proto-cuneiform to the end of the Mesopotamian dynasties 3,000 years later, Mesopotamians 'signed' documents by rolling a carved cylindrical seal across the soft clay. Seals were small – typically no bigger than a bottle cork but often much smaller – and made of a durable material like stone or a more precious substitute. They were often hollow so that they could be worn on a cord around the neck. Seals would be carved with intricate figures and symbols depicting anything from deities and mythological creatures to heroic deeds and scenes of daily life. They were often works of art in themselves, and became a defining object in Mesopotamian identity and culture. Everyone from the emperor to the lowliest slave had one, for 'signing' contracts, establishing ownership, and – in a society where there was no distinction between religion, magic and 'reality' – warding off evil spirits.

Mesopotamia, c. 3500 BCE

Proto-cuneiform spread rapidly to the surrounding regions and enabled huge societal changes. Having access to the means to record information on tablets gave the priests of Mesopotamia an information storage and retrieval system independent of human memory, and of almost infinite capacity. Having administrative systems allowed towns to grow much larger than they had ever been, and eventually become city states.

As cities grew they became more and more stratified. Priests and priestesses became more powerful, and eventually were taken over by local leaders or 'strongmen', who provided protection for the cities from raiding neighbours. These strongmen became the first kings, and they raised armies, waged wars, and built great palaces for themselves and temples to the gods, culminating in the great ziggurats that still exist today.

All was administered by records kept on small clay tablets. They were simply records, with no value in themselves once a transaction was complete. When archaeologists excavated the site of Uruk in the twentieth century, they found thousands of them on a rubbish heap, discarded like a pile of old till receipts found at the bottom of a handbag.

With around 1,200 different signs, proto-cuneiform was complex. Reading and writing it became a highly specialised skill. The only people who knew the system were scribes who trained for several years in scribe schools. They worked on the clay equivalent of exercise books, practising signs and symbols over and over, and doing spelling and vocabulary exercises.

The proto-cuneiform system didn't attempt to represent language directly in the way that we think of reading and writing today.[5] The tablets contained detailed lists, but did not have sentences or a grammatical structure, and weren't linked to a particular spoken language. This limited its scope but it also made it easier for proto-cuneiform to spread from Uruk to the surrounding regions.

By creating a shared set of visual symbols representing objects, times and places, the scribes began the process of outsourcing memory. Clay tablets could capture and store details no person could be expected to remember. They also dealt the first blow to two of our other senses: tactile, three-dimensional tokens were replaced with visual symbols and oral communication was, for the first time, downgraded in favour of written records. Visual communication was on the rise.

Seeing Language

There is no evidence of proto-cuneiform being used for anything other than administration. All the other knowledge within the Mesopotamian culture – its stories, beliefs, incantations, recipes, songs and so on – remained within the oral firmament.

Then, around 3000 BCE, a dramatic innovation took place. Someone – presumably a scribe – started using pictures and symbols to form spoken words in writing. Words began to be constructed using signs with meanings that sounded similar to the corresponding part of the word, as in a game of charades when a player acts out a word by portraying words that sound like each of its syllables, one by one. Scholars call this type of writing a syllabary.

This was the beginning of phonetic writing: an enormous inventive leap. In phonetic writing the written symbol represents not an object but a word describing an object, making it an abstraction of an abstraction. The first known phonetically produced written word is the name of an early king, found engraved on a gold bowl from around 2700 BCE. The four syllables of his name – MES-KA-LAM-DUG – are depicted by four characters; the meaning of each character sounds like the corresponding syllable.

In this new writing system, called cuneiform from the Latin for wedge-shaped, the number of characters used fell to around 600 and the signs themselves became more and more abstract as scribes adopted conventions to make writing easier and faster. Cuneiform writing was adopted for several different languages spoken around the region, each of which modified it to suit their own purposes.

Once cuneiform represented spoken language rather than simply facts and figures, it could record anything that could be expressed in words. Literature, religion, law, medicine, magic, as well as all the mundane day-to-day communications that occur between people could be captured in a permanent, visual form.

Various forms of cuneiform writing, in the hands of countless, nameless scribes, recorded the comings and goings of the people of Mesopotamia for 2,500 years. The first great world empires – Akkadian, Assyrian, Babylonian and neo-Babylonian, Persian – were established under cuneiform and the first great work of literature – *The Epic of Gilgamesh* – was recorded. Laws were written down for the first time, most famously in the Code of Hammurabi, in around 1750 BCE. Cuneiform recorded the names and great deeds of kings, the glory and wrath of gods, the structure of society, and the legends and laws of the land.

The Sumerians and their successors grasped the power of writing to capture, cement and even create history. The Sumerian King Lists were carved lists of the names, reigns and places of their kings and kingships (among whom is a lone queen), and those of neighbouring dynasties, going back as far as memory served, ascribing some of their early kings a somewhat epic lifespan stretching into tens of thousands of years. The lists divided past rulers into those before and those after the 'great flood', causing great excitement when Victorian scholars first discovered them. It turns out, as modern scholars have since discovered, that the idea of a great deluge is a motif that appears in the captured oral histories of traditional cultures all over the world, not just in the biblical Noah's ark story.[6]

Assyria, *c.* 640 BCE

King Ashurbanipal, Emperor of Assyria, was one of most powerful men in the history of the world. He ruled his empire with ruthless efficiency, crushing not only his enemies but also their ancestors, descendants and conspirators. A contemporary relief sculpture shows him relaxing in his garden while the severed head of an enemy hangs on a nearby tree.[7] Ashurbanipal was also one of the only literate kings. Throughout the long history of cuneiform, reading and writing remained the almost exclusive preserve of scribes, but Ashurbanipal enjoyed his reputation for scholarship so much he ordered his court artists to depict him with a stylus – a tool for writing on clay – as well as a sword.

Ashurbanipal sent scribes and scouts out around the entire known world collecting knowledge. Everything they found was transcribed into tens of thousands of cuneiform tablets and stored in the great library he built at Nineveh, near modern Mosul in northern Iraq. Tablets were carefully filed under categories including history, government, religion, astronomy, magic and so on. There was even a locked section that stored state secrets away from prying eyes.

The library at Nineveh was buried by raiders not long after Ashurbanipal's death in 627 BCE, and the Assyrian Empire came to an end soon after. The sum total of human knowledge at the time, stored in the library, was forgotten for 2,000 years until Henry Rawlinson's friend, Austen Henry Layard, discovered it in the 1850s. Even then it was several more decades before the contents of the library could be deciphered and their extraordinary intellectual bounty explored.

Persia, *c.* 521 BCE

Darius I became Emperor of Persia after slaying Gaumata, the great pretender, who had taken the throne by impersonating the brother of the preceding ruler. The first few years of Darius' rule were characterised by a series of revolts and insurgencies in the provinces throughout the empire, which Darius ruthlessly and successfully put down. Once his rule was secure, he set about proclaiming his glory, his royal lineage and his favour with the gods. As befitted his status, Darius commissioned a new version of cuneiform writing, then sent out huge teams of scribes, artisans and stonemasons to create monuments to him around the empire. He ordered that each monument must comprise a sculpture depicting his victory over Gaumata and the twelve rebel lords, watched over by the god Mazda, and an inscription relating Darius' lineage and the story of his accession. The inscription was to be written in three scripts, the new text (now called Old Persian by modern scholars), and the traditional Babylonian and Elamite. The largest monument of all was to face the Royal Road, on the great mountain of Bagastana (Behistun) in the Zagros, where all who passed by would be reminded of Darius' greatness forever more.

The crew worked painstakingly at Behistun for several years, first smoothing and preparing the cliff face, then painstakingly carving the sculpture and its accompanying inscription, sign by sign, into the stone. Where they made an error they repaired it using lead, making sure each character's form was perfect. At some point during the project they realised they hadn't left enough room for one of the inscriptions, so they began it all over again in a different place.

When they finally finished they cut away parts of the wall below to make the inscription inaccessible, to prevent damage from enemies or later rulers. The result was magnificent, and could be seen for miles around, although no one could read the inscription itself, even from close up: it was too high to be legible. No matter, as so few people could read anyway. And besides, the monument was certainly seen from the road – it was noted by European travellers as early as 1598. The stonemasons' defences were effective, and no one managed to scale the cliff face for more than 2,000 years, until 1836, when a cocky English gentleman soldier in search of his own form of glory decided he wanted to copy the script.

The Persian Empire continued for 150 years after the reign of Darius, until 336 BCE, when the Greek warrior Alexander the Great brought 3,000 years of Mesopotamian empires to an end. Cuneiform was no longer the script of rulers, and it fell into disuse. Within a few centuries the world's first writing system was completely forgotten, along with the records of the world's earliest civilisations, and the beginning of what we call history. It remained so for nearly 2,000 years.

The Spread of Writing in the Ancient World

As cuneiform was developing in Mesopotamia, the Egyptians took the concept of a sound-based, visual representation of language and created their own version, later named hieroglyphics (from the Greek for 'sacred carving').[8] Like early cuneiform, hieroglyphics were a mixture of phonetic sound signs, symbols that represented whole words, and signs that contextualised words, and had a similar number of characters, around 600. However, writing seems to have played a very different role in Egyptian life than in Mesopotamia.

Unlike their neighbours to the east, Egyptians didn't use hieroglyphs to manage the Egyptian economy. Rather, they used them to glorify and ensure the eternal life of their pharaohs. Hence the script was closely associated with magical power. The Egyptians believed that preserving the written name of someone preserved their soul. To this end, every noble commissioned their own 'stela', a great stone column engraved with their name and text commemorating their life and achievements. Egyptian writing was also closely linked to its artistic traditions, preoccupied with religion and the afterlife, and hieroglyphics often appeared alongside paintings and sculpture.

The wedge-shaped appearance of cuneiform reflected the medium on which it was written: the plentiful clay of the Fertile Crescent. By contrast, the Nile Valley was plentiful in papyrus, a tall, thin-stemmed marsh plant. Egyptians created a writing surface by peeling the inner stems of the papyrus plant into strips and laying them side by side, then putting another layer perpendicularly on top of the first and beating them together into a pulp. This created a strong, smooth, flexible sheet for writing on that could be joined to other sheets to form a scroll. This new product proved hugely successful, and papyrus remained in use as a writing medium for several thousand years, until it was succeeded by parchment and, in the Middle Ages, by the Chinese invention of paper.

In Egypt, literacy was the privilege of the ruling classes of priests and nobles rather than an administrative class of scribes.

As such, literacy was more common than in Mesopotamia, but still well beyond the reach of common people. Writing controlled the flow of information and thus was closely linked to power, both spiritual and secular, and was a precious and jealously guarded skill.

Egypt, France, England, early 1800s

Twelve years before Henry Rawlinson started recording the Behistun cuneiform inscriptions, a famous breakthrough had been made in deciphering Egyptian hieroglyphics. In a saga of imperial rivalry and pride, the secrets of this other long-lost writing system were revealed by a stone. At the height of the Napoleonic Wars in 1799, French soldiers stationed in the Egyptian port town of Rosetta found an unusual stone built into a wall. Recognising its potential, the French removed and stored it with the huge haul of Egyptian antiquities they were assembling. Two years later the British defeated the French in Egypt and wasted no time in demanding the French collection of antiquities as war booty. The French general in charge refused to hand them over and hid the stone, considering it his personal property. In mysterious circumstances, the British managed to 'retrieve' it — in these colonial times the Egyptian authorities were not involved at any stage — and shipped it straight back to London, where it was put on public display as a symbol of Britain's victory. It remains to this day one of the most visited objects in the British Museum, despite vociferous recent demands from the head of the Egyptian Supreme Council of Antiquities for its return.[9]

What made the Rosetta Stone so precious was that it contained three inscriptions, in classical Greek, Egyptian hieroglyphs and an unknown script believed to represent the

Coptic language. Scholars were familiar with the Greek text, and it told them that the three inscriptions were identical. This implied that the Stone could provide the key to deciphering the other two texts – most enticingly, the hieroglyphs.

For centuries it had been assumed that the stylised pictorial markings in hieroglyphics were pictograms. That is, they symbolised abstract and allegorical ideas associated with the object depicted. It was thought, for example, that the hawk symbol represented the idea of speed, a crocodile meant evil, and a falcon denoted victory and divinity. This assumption came from a fifth-century Greek scholar named Horapollo, who wrote a text called *Hieroglyphia* purporting to explain the enigmatic script. The book was discovered and circulated widely in Europe in the fifteenth century, and while Horapollo's analysis turned out to be completely wrong, his ideas were deeply entrenched among European scholars by the time the Rosetta Stone was discovered.

The idea of a symbolic language is seductive, and especially plausible in light of the many other exotic aspects of ancient Egyptian culture. However, an unknown symbolic text has an almost infinite number of potential interpretations. The Rosetta Stone promised a direct route to translation and a huge step towards understanding this mysterious writing system and probing the recondite Egyptian culture.

The race to decipher hieroglyphs took place in the wake of the Duke of Wellington's final victory over Napoleon at the 1815 Battle of Waterloo. After decades of armed conflict between England and France stretching back to the 1750s, nationalist sentiments were running high on both sides of the Channel. On the English side was a doctor named Thomas Young, a polymath later described by his biographer as 'the last man who knew everything'.[10] In 1815 he was 42 and had published ideas as diverse (and correct) as a wave theory of light that contradicted Newton, and the hypothesis that the retina of the eye has three colour receptors. He was fluent in many languages and a meticulous draftsman, and he became interested in the Rosetta Stone when he read that the unknown Demotic script might comprise an alphabetic writing system; that is, a system of writing in which symbols represent a single sound rather than a syllable.[11]

Young made meticulous copies of the Rosetta inscriptions. He noticed that some of the Demotic signs looked similar to certain hieroglyphs and suggested that Demotic might be a cursive version of the more formal hieroglyphs, analogous to the modern difference between handwriting and printing. He also proposed that Demotic was not a full alphabet but a combination of alphabetic and pictographic signs. These were to prove critical insights.

Young focused on the Stone's six hieroglyphic cartouches. These were the encircled symbols known to contain royal names. He guessed that foreign names would be written phonetically and, progressing on this basis, eventually proposed thirteen phonetic hieroglyphic 'letters', six of which proved to be correct. He published his findings in 1819.[12]

Meanwhile, across the Channel, another man was working on deciphering the hieroglyphs. Jean-Jacques Champollion was a young scholar obsessed with the Coptic language who was determined to be the first to solve the hieroglyphic puzzle. With some prescience, the French authorities had made paper copies of the Rosetta's inscriptions before the Stone was taken from them, and had also gathered many other copies of hieroglyphic texts while they were in Egypt. So while he didn't have access to the Stone itself, Champollion had plenty of hieroglyphic material to work with.

Like Young, Champollion focused his attention on the cartouches. He started with cartouches of Ptolemy and Cleopatra, both of Greek origin, and managed to demonstrate that their five overlapping letters – p, l, o, e and t – corresponded to five identical hieroglyphs. Champollion applied the same approach to other foreign names and ended up with around forty hieroglyphic sound signs. He then proved the same approach worked with Egyptian names, demonstrating that phonetic hieroglyphs were not limited to foreign words. With these insights Champollion demonstrated that the hieroglyphic writing system was a sound-based system and not, as had been believed for centuries, symbolic in nature.

Champollion presented his discoveries publicly in September 1822, without crediting Thomas Young in any way. A debate ensued among scholars about the relative merits and importance of each party's contribution, drawn largely along national lines. The debate persists in some scholarly quarters to this day, but outside academic circles it is without doubt Champollion who is credited with deciphering hieroglyphics. It was his celebrity that Rawlinson craved when he wrote in a letter to his sister a few months after making his first copies at Behistun: 'I aspire to do for the cuneiform alphabet what Champollion has done for the hieroglyphics.'[13]

Eastern Mediterranean, *c.* 1000–500 BCE

On the eastern shore of the Mediterranean between Mesopotamia to the east and Egypt to the south, a great seafaring civilisation grew. More a chain of independent city states sharing a common language than a nation, they originated in Byblos in modern-day Lebanon and spread west around the Mediterranean coast, establishing trading posts and colonies as far away as Carthage in modern Tunisia and Cádiz in Spain. Scholars disagree about whether they had a name for themselves[14] but the Greeks called them the Phoenicians, from their word for dark red,[15] referring to a precious purple dye in which they traded, or possibly to their ruddy complexions. They spoke a version of the Semitic language common to a wide racial group that lived all over the Near East throughout antiquity. Writing became an important administrative tool for these commercially minded people, but as a practical, pragmatic bunch they simplified the Egyptian system substantially. Around 1050 BCE they stripped hieroglyphics of all signs that represented whole words, leaving just twenty-two signs that each represented a syllable or consonant sound. There were no signs for vowel sounds, which were expected to be known or inferred by the reader. This was the world's first alphabet, the lack of vowels making it a particular form scholars call an adjab.[16] The Phoenician adjab was to become the basis of all the major Western alphabets including Greek, Roman, Hebrew, Arabic, Aramaic, Germanic, Cyrillic and Coptic.

By 900 BCE the Near East was a connected network of civilisations. Trade routes linked cities along the Tigris and Euphrates with Anatolia (Turkey) and the Mediterranean, where shipping routes carried goods and people to Egypt, Cyprus, Greece and Spain. Greece had used a form of writing known as Linear B for administrative purposes during the rule of the Mycenaeans, but this had fallen into disuse when the Mycenean Empire fell around 1200 BCE, leading to a 400-year period known as the Greek dark ages.

One important aspect of Greek culture that was sustained through the dark ages was its tradition of spoken poetry. Ancient Greek poetry had a very particular form and rhythm, with many repetitions and familiar phrases. In particular, it employed a metre with one long and two short vowel sounds. It was typically devised on the spot by *aoidos*, or bards, and performed with musical accompaniment. The most famous of the bards was

Homer, reputedly a blind man from Ionia, although the facts of his background have never been firmly established.

Although some scholars disagree, one prominent school of thought has it that it was the desire to capture Homer's poetry for posterity that brought about the Greek alphabet, which in turn spawned the Roman alphabet we use today.[17] According to American scholar Barry Powell, around 800 BCE an individual of Homer's acquaintance, someone who knew the Phoenician script with its twenty-two characters, adapted the system to capture the sound of language more precisely. He reduced the phonetic representation (sound) of each Phoenician symbol from a syllable to the smallest possible unit, the consonant, then added several new signs to represent vowel sounds. This created a truly phonetic – from the Greek word *phone*, meaning sound, and not to reflect its Phoenician origin – writing system, wherein pronunciation was linked directly to the letters without requiring knowledge of the underlying language. With this new system the inventor could capture Homer's poems exactly as they had been spoken in a permanent, visual form.

Thus came about the first true alphabet, capable of representing speech exactly as it sounds. Powell supposes that the inventor of the Greek alphabet spent many hours with Homer as he dictated his epic poems – *The Iliad* alone takes twenty-seven hours to perform in full, as Powell points out. Homer could take advantage of the writer's relatively slow pace to perfect and embellish his usual repertoire.

Once transcribed, copies of the poems could be made and circulated, presumably alongside instructions on how to decode the alphabetic technique. People quickly realised the possibilities the new alphabet afforded, and it was widely adopted throughout the Eastern Mediterranean.

Powell cites as evidence of the 'orality' theory the earliest known examples of Greek writing, typically fragments scratched onto pottery shards or into stone; ironically, more substantial documents were probably written on papyrus or parchment and would have long disintegrated, leaving only 'broken shadows from a reality otherwise invisible'.[18] What fragments exists are almost all written in the metrical style of Greek literature, and are exclusively poetic or comedic graffiti – jokes and puns clearly linked to spoken language. There are no examples of early Greek used for administrative purposes; the opposite is the case for the earlier Greek language, Linear B, which was purely administrative.

Alongside the rise of the alphabet – many scholars claim because of it – Greek politics and culture soared over the next few hundred years. Poetry, philosophy, art, sculpture, architecture, mathematics, science all flourished in the wake of the simple, twenty-six-letter Greek alphabet. Its simplicity made it widely available to the general population and must surely have influenced the development of democracy in the rising city states of Athens and her sisters.

The Greek civilisation that arose in the years following the invention of its alphabet laid the foundation for Western culture up to and including the present day. With a simple, flexible tool to formulate, capture and express ideas, and a relatively easy means to distribute them, the Greeks became the oracle. The first few centuries of the Greek alphabet produced a succession of thinkers – Pythagoras (b. 568 BCE), Sophocles (b. 490 BCE), Herodotus (b. 484 BCE), Euripides (b. 480 BCE), Socrates (b. c.470 BCE), Hippocrates (b. 460 BCE), Aristophanes (b. 446 BCE), Democritus (b. c.460 BCE), Plato (b. 427 BCE), Aristotle (b.322 BCE), Euclid (b. c.300 BCE), Archimedes (b. c.287 BCE) and others – who bequeathed the first versions of most of the subjects studied in schools and universities around the world ever since: philosophy, medicine, geometry, history, the sciences, literature, and politics.

Bagastana, 1836

As Henry Rawlinson toiled away on the narrow ledge, copying the intricate inscriptions and dreaming of glory, others were pursuing the same quest from the relative comfort of Europe. Working from copies of inscriptions made by earlier travellers to the Near East, scholars in Germany, Denmark and France were making slow but significant progress toward deciphering cuneiform.

The challenge Rawlinson and his rivals faced was far greater than that presented by hieroglyphics twenty years earlier. Unlike the Egyptologists who had the Rosetta Stone's familiar Greek script as a starting point, none of the three cuneiform scripts found at Behistun and elsewhere in Persia were familiar, and neither were the underlying languages. They had all been forgotten for nearly two millennia.

Deciphering cuneiform required two stages. First, the cuneiform signs needed to be converted into a familiar alphabet, in a process known as transliteration. Secondly, the transliterated words needed to be translated into English. It is almost impossible to imagine doing either of these entirely from scratch, but some inroads had been made by European scholars in the decades leading up to 1836.

Scholars had determined that the inscriptions represented three different languages, and that the simplest text – Old Persian, the one that Rawlinson was copying at Behistun – comprised fewer than forty different signs, making it the most promising starting point. Links had also been established between this text and the ancient language of Zoroastrianism, which had been partially translated into French in 1771.

In 1802, a German scholar named Grotefend concluded that Old Persian text was phonetic and identified several letters by seeking out the names of kings and common phrases known from later Greek inscriptions, such as 'king of kings'. Earlier in 1836, two French scholars claimed within a month of one another to have made substantial improvements on Grotefend's work and, unbeknown to Rawlinson, these new alphabets were being prepared for publication even as he was making his copies at Behistun.

Meanwhile, Rawlinson spent the late summer analysing the copies he'd made. He managed to construct an Old Persian alphabet, with phonetic values assigned to eighteen characters. He was confident his solutions were superior to Grotefend's. He had no idea of the recent French decipherments at this stage. Back in Europe, neither Rawlinson's efforts nor even his existence were known to his Orientalist would-be colleagues and rivals.

In early 1837 he returned many times to Behistun to copy more text. He managed to complete more than 200 lines of Old Persian – by a long way the most available to anyone – when military duties again intervened.[19]

Later in the year, Rawlinson finished translating the first few paragraphs of the Old Persian inscription, and on the first day of 1838 he sent a transcription and translation of the first two paragraphs to the Royal Asiatic Society. This was his first contact with the scholarly community of Orientalists back in Europe.

The two paragraphs he sent were:

[1.1] I am Darius, the great king, the king of kings, the king in Persia, the king of countries, the son of Hystaspes, the grandson of Arsames, the Achaemenide.
[1.2] Says Darius the king: My father is Hystaspes, the father of Hystaspes is Arsames, the father of Arsames is Ariaramnes, the father of Ariaramnes is Teispes, the father of Teispes is Achaemenes.

In a long letter accompanying the translations, Rawlinson described the saga contained in the inscription: the story of the ruler King Darius the Great. His communication caused great excitement in Europe. Not only did this unknown young man appear to have made a major breakthrough in deciphering cuneiform, but the details of the Darius story he related were consistent with the writings of the classical Greek historian Herodotus (b. *c*.484 BCE).

At this time the only sources of information about the ancient Near East were the books of the Old Testament and Herodotus' book *The Histories*. Historians had no idea how accurate these accounts were. Rawlinson's tiny piece of Old Persian translation and description of the story suggested that Herodotus was indeed a reliable source of historical fact, and that Behistun may provide a first-hand account of events that took place twenty-five centuries earlier.

European scholars gave Rawlinson a warm response, but his distance from the active world of scholarship and his ongoing military duties proved a significant impediment to his dreams of intellectual triumph. He didn't receive the new French translation until June 1838, by which time he recognised that he had been pre-empted.

Nevertheless, he continued his studies, spending months preparing a detailed 'Memoir on the Persian Cuneiform Inscription at Behistun'. In it he claimed to be the first to fully decipher Old Persian, but just as he was about to send it off to London, Rawlinson received a letter from a Norwegian scholar, Christian Larssen, saying he had translated the entire Old Persian alphabet. Rawlinson had been pre-empted again.

Again he decided to carry on, and was in the process of preparing a full translation of the Old Persian Behistun inscriptions, with extensive notes and commentary, when he was called away again, this time to Afghanistan in the lead up to the First Anglo-Afghan War. It was to be another six years before he could return to cuneiform.

The Impact of Writing

The written alphabet translates the aural, invisible elements of language into graphic, visible signs.[20] But writing involves more than just the visual capture of language. It is a new form of expression of human thought. Writing makes possible levels of abstraction and complexity that would be impossible without it. A writer has the opportunity to organise thoughts and find ways to convey them to a reader that almost nobody could do verbally, unless performing a rehearsed text. In a very real and practical sense, writing expands the potential of the human brain.

It is impossible for most of us to imagine a world without reading and writing. Once learned, it quickly becomes such a fluent form of communication that it feels like one of our senses and, as we've seen, human brains have an area dedicated to recognising text. I recall my 5-year-old nephew telling me that, now he could read, he couldn't help reading everywhere he went. And that's how it is for most of us. If you've ever been in a place where you don't recognise the local script, you will probably have felt dislocated, as if one of your senses has been shut down.

But reading and writing are not natural; they are man-made inventions. While most children will naturally learn to speak without active teaching, reading and writing must be taught and learned formally. We are hard wired for oral communication, but we must programme in its visual representation.

But what an invention! What a liberation from the limitations of our brains! In its first incarnation in Mesopotamia, writing was an information storage and retrieval system. It allowed scribes to tally up offerings to gods and goddesses, and later homages and taxes. Without such a system of capturing and storing information – far beyond the capacity of any human brain – the early cities could not have grown and prospered, nor built the fabulous monuments to their kings and gods that they did, nor raised armies and conquered their neighbours, nor established vast empires across the lands between Europe and Asia. This was writing as additional storage space for the brain – not particularly exciting in itself, but extremely useful to those in whose service it was deployed, and a great, if subtle, enabler of change.

Once writing could express spoken language, its potential became almost infinite. Any thought, idea, description, sentiment, proclamation, or simple exchange of pleasantries could be passed on, to an individual, to some, to many, or even kept for the writer alone. These expressions could be carefully constructed, revised, edited, embellished, and made as perfect an expression of the writer's mind as the language and the writer's capabilities allowed. They could then be analysed, critiqued and shared by others near and far.

I love the idea that our own Roman alphabet – direct descendant of the Greek – may owe its vowels to poetry. A whole book could be written about the delights of the written word, but thousands already have been, and I need only say that literature must surely be one of humanity's greatest achievements.

It is yet another great irony in our story, however, that the very alphabet that was invented to capture the wonder of Homer's spoken poetry became so successful that it killed most oral traditions including, in most parts of the world, oral verse.

Beyond storing large dumps of information outside our brains, writing vastly increases our brains' processing power. Most people can't figure out complex numerical problems beyond simple arithmetic in their heads – hence the mathematician's blackboard, covered in equations, or the scribbly long division on the back of a shared restaurant bill. Having a visual representation of one's thoughts – the starting assumptions, the possible logical progressions – and the ability to go backwards as well as forwards, are the very business of serious thought across every imaginable discipline.

While many bemoan the decline of the handwritten letter, the fact is that today's world is more writing-driven than ever. Even if they don't demand full sentences, Twitter, texts, emails, chats and most of the other contemporary communication media are writing-based.

Western historians denote the invention of writing as the beginning of history. Some modern scholars and commentators deride the traditional view of written history as a record of the comings and goings of kings and wars and empires. Less obvious is the fact that it was the invention of writing that allowed kings and wars and empires to exist in the first place.

While writing unlocks the potential for the exploration of complex ideas and rational thought, the inherently private, individual nature of reading and writing also changed the nature of learning and memory forever. Plato identified this risk 2,500 years ago when in *Phaedrus* he told the story of the god Theuth, inventor of numbers, calculation, geometry, astronomy and writing. Theuth visited the Egyptian King Thamus and showed him these inventions so that he could make them widely known and available to the Egyptian people. When it came to writing Theuth declared, 'Here is an accomplishment, my lord the King, that will make the Egyptians wiser and give them better memories; it is a specific both for the memory and for the wit.'

Thamus pointed out that the parent of an art is not always the best judge of its utility or otherwise to its users, and went on:

This discovery of yours will create forgetfulness in the learners' souls, because they will not use their memories; they will trust to the external written characters and not remember of themselves. The specific which you have discovered is an aid not to memory, but to reminiscence, and you give your disciples not truth, but only the semblance of truth; they will be hearers of many things and will have learned nothing; they will appear to be omniscient and will generally know nothing; they will be tiresome company, having the show of wisdom without the reality.[21]

Plato feared that writing would replace active, face-to-face debate, with the result that true understanding – only achievable, he believed, through an interactive process of interrogation and discussion – would be lost.

Certainly writing changed some fundamental aspects of daily life and culture. Writing captured myths and legends, which for millennia had been shared in spoken form, never told exactly the same way twice but honoured in spirit. Once written down they often became doctrine, fixed and unyielding. The natural fluidity of the oral tradition was replaced with rigidity: traditions become dogma, norms were replaced with rules and laws, pragmatism with judgement. Things that would naturally be forgotten in oral cultures were remembered, and much of what would previously have been remembered was consigned to the written word and gradually forgotten.

Postscript: Sir Henry Rawlinson, 1st Baronet, 1810–96

Henry Creswicke Rawlinson was born at the height of the Napoleonic Empire's domination of Europe. He dedicated his adult life to the service of the British Empire, and spent two decades trying to unlock the secrets of another empire created 4,000 years earlier. He combined a keen classical intellect with a love of action, physical bravery, and not a little charm. His successful army career and the esteem in which he was held by local people – from the Shah of Persia to his local troops to Arab mountain tribesmen notorious for murderous inclinations – suggest his diplomatic skills were considerable. His letters home spoke of regrets over drinking sessions, gambling debts and foolish pranks, suggesting a well-developed sense of fun, while his cuneiform pursuits indicate a rich intellectual hinterland. He was physically competitive, proficient and courageous; on one engagement Rawlinson rode for 150 straight hours, covering 750 miles, to deliver an urgent message. If he wasn't the inspiration for Indiana Jones, he should have been.

Rawlinson's upbringing was genteel but not wealthy. He was born in 1810 and grew up on an idyllic Oxfordshire estate in a Cotswold stone manor house surrounded by lawns, meadows and fields. His father spent his days riding, hunting and shooting with the local aristocracy. Rawlinson inherited his father's taste for an English gentleman's pursuits and displayed some considerable talent in these. Even as a lad he was a crack shot and an excellent horseman. However, he was the second son, so when money became tight economies were made with his education. He left school at 16, already 6ft tall, and joined the army of the British East India Company, the enormous private company set up by Queen Elizabeth I to establish trade with the East Indies. By 1826, the 'John' Company, as it was known, was running the British Empire from the Near East to China. For an ambitious young man of limited means the Company offered the opportunity of adventure, advancement and, possibly, fortune.

As Henry Rawlinson climbed to the ledge below the Behistun inscriptions for the first time in 1836, intent on deciphering cuneiform, he had no idea what historical treasures that decipherment would eventually reveal. The great library of King Ashurbanipal and its 30,000 clay tablets were still buried under the desert and would not be discovered for another twenty years. The granite stela proclaiming the Code of Hammurabi – a system of laws and values strikingly similar to those in the Bible, yet written centuries earlier – would remain undiscovered in Rawlinson's lifetime. The city state of Uruk and its thousands of proto-cuneiform tablets and their clues to the very origins of writing would remain undisturbed for close to another century.

Despite the disappointments of his first contact with the Orientalist community back in Europe, Rawlinson persevered with his efforts. He returned to the rock face at Behistun many times and eventually managed to make copies of the entire inscription, including all three forms of cuneiform. He was eventually successful, with others, in deciphering the Old Persian and also the Babylonian texts, which unlocked most of the archaeological treasures of ancient Mesopotamia. Although it cannot be said that Rawlinson alone deciphered the cuneiform texts, he was undoubtedly more than instrumental in the process.

Rawlinson returned to England for the first time in 1847, twenty-two years after leaving as a cadet. To what must have been his great delight, he was received as hero in the capital. He was presented at court and dined with the queen, was awarded various honours and fellowships from societies and universities all over the world, and was reportedly the toast of London society. He presented a paper at a meeting of the Royal Asiatic Society chaired by Albert the prince consort. Many years later, in his 1895 obituary in *The Athenaeum*, he was given the 'credit of having contributed more than any other man to the unravelling of the Persian cuneiform by his laborious and scholarly publication of the great inscription of Darius at Behistun'.[22]

The cuneiform writing system began the process of capturing human history, and at the same time instigated history as we know it. Writing translated the multisensory, three-dimensional world into a two-dimensional visual record, creating a peephole into the future. It was also the writing on the wall for oral traditions. The Greeks who created the alphabet – the ultimate writing system – and created philosophy and democracy, also spawned warriors and tyrants. In the 330s BCE, Alexander, King of Macedon and Greece, conquered Persia and Egypt and established the Greek alphabet as the official writing system. Within decades cuneiform and hieroglyphics fell into disuse and were forgotten. The peephole closed.

When Champollion, Rawlinson and others deciphered hieroglyphics and cuneiform, two millennia later, they reopened the peephole to unlock the stories of the world's first civilisations. Their work allowed us to see through time, to understand how and where our world came from, and ponder again: how did we get here?

BELIEVING

WHEN WE DIDN'T SEE

7

AMONGST BARBARIANS: THE AGE OF THE INVISIBLE

> I dwell amongst barbarians, a proselyte and an exile,
> for the love of God. He will testify that it is so.
>
> St Patrick (385–431), Epistle to Coroticus

Among the priceless collection of early English manuscripts held by the British Library is one known as Harley MS 585, written by anonymous scribes in the tenth or eleventh centuries.[1] It is known as the *Lacnunga* and contains a miscellaneous collection of remedies, charms and prayers written in Old English and Latin. Among them is the Nine Herbs Charm, a treatment against poison and infection. The charm requires nine herbs: mugwort, dockleaf, watercress, chamomile, nettle, one other whose translation is not certain, crab apple, chervil and fennel. The healer must pound the first six herbs into dust, singing an incantation that describes each plant's qualities three times over. When the first six herbs have been crushed, a serpent will appear crawling along the floor. This is the spirit of the infection that the healer must destroy. To do this the healer must re-enact a legendary feat of the Norse God Woden using nine wooden sticks with the initial rune of each herb's name cut into them. When the serpent appears, the healer and the sticks are imbued with the power of Woden, and when the healer strikes the approaching serpent with the sticks it shatters into nine pieces. That done, the healer must mix the pounded herbs into a paste with ash, water, a beaten egg and the juice of the crab apple. Then the healer adds the last two herbs, chervil and fennel – the most powerful herbs, the charm says, and created for all mankind by the Lord while he was hanging – to create a poultice. The healer must then sing the charm into the mouth of the patient, into both their ears, and over their wound, before finally applying the poultice. This remedy will now provide strength against enemies, flying creatures, enchantments, poisons of all colours and kinds, wounds and blisters. Christ, it says, stood over disease of all kind.[2]

The Nine Herbs Charm was written in Old English in the verse form typical of Anglo-Saxon poetry. It is in part a practical herbal recipe and in part a magical incantation. It refers both to Jesus Christ and to the pagan Norse god Woden, while the phrase 'hanging Lord' could refer either to Jesus on the Cross or to Woden, who in Norse lore suffered hanging from a tree for nine days and nights. Its claims are part science, part magic, part miracle. No distinction is made between its material and supernatural elements: the invisible stands on equal terms with the visible.

The same manuscript also contains the Charm Against a Sudden Stitch, a remedy against attack by elves. Any sudden, unexplained pain was believed to be caused by 'elf shot' – an invisible arrow hurled by elves and lodged into the site of the pain. The charm removes the arrowhead. Malicious elves, along with serpents, dragons, Valkyries and goblins, were just some of the unseen actors playing their part in the everyday lives of early medieval people.

The Nine Herbs Charm was written down in the middle of a time, lasting almost 1,000 years, when believing trumped seeing.

Western Europe, *c.* 300–500 CE

In the eighth century BCE, just a few decades after an anonymous Greek scribe was inventing the alphabet, a great city rose up in the Italian peninsula. Founded by Romulus, the twin abandoned with his brother Remus as an infant and famously suckled by a she-wolf, Rome grew to become the greatest empire the world had ever known. For more than 1,000 years Rome ruled a civilisation that, at its peak, stretched from Britain (Britannia) to North Africa in the west, around the entire Mediterranean Sea both north and south, and as far east as Syria.

By the third century of the current era (CE) the Roman Empire was struggling under pressures of invasion, civil war, natural disasters and plague. In 284 CE it divided into two parts, west and east, but this didn't solve its problems. Over the next two centuries Germanic tribes from the north and east made deeper and deeper incursions into the Western Empire: Vandals into Northern Africa, Visigoths into Spain, Ostrogoths into Italy, Franks into Gaul, Huns into Western Germany, and the Angles and Saxons into Britannia. Romans were driven further and further east until, with the final defeat of the Emperor Romulus in 476 CE, the Western Roman Empire was no more.

As the Romans retreated, indigenous populations left behind had little defence against the Germans. They were quickly defeated and colonised by the invaders, who imposed a very different regime from the Romans. Although some

German tribes had lived close to the Romans for many years and adopted their ways, those from farther afield were a people completely apart. As the Roman historian Tacitus wrote in the first century CE, Romans lived in elegant cities or country estates in houses built of cut stone and decorated with paintings, mosaics and sculpture. They wore fine, fashionable clothes and prized erudition, literature, history and high aesthetic sensibilities. They operated sophisticated public works and infrastructure and efficient bureaucracies and legal systems. The Germans, by contrast, lived in scattered, makeshift huts erected from whatever materials were to hand. They wore crude robes made from a rough fabric or animal skin, fastened with a pin and left naked underneath. They had little regard for possessions other than their treasured weapons, which they carried at all times, including to the table and, reputedly, to bed. According to Tacitus, even these weapons showed little ostentation and minimal decoration. It is not difficult to see why the Romans described the Germans as barbarians.

Tacitus' description was naturally somewhat biased. In fact, the German tribes had well-developed cultures, but these were a world away from the sophisticated Roman way of life. The Germans valued strength and loyalty above all, and this was reflected in their social structures. Each tribe had a chief or king, determined by nobility, and a number of princes who attained their position through birth or by great deeds. Princes were attended by followers who competed for the prince's favour, while princes competed among themselves for followers. Followers protected their prince at all costs, while the prince relied on waging war to sustain the enthusiasm of his entourage and acquire plunder with which to equip them with weapons and supply them with daily meals and drink. This dependence on plunder provoked a constant cycle of raid and counter-raid, interspersed with bouts of drunken feasting and lazing around. As Tacitus put it, the Germans found it 'stupid and spiritless … to acquire by their sweat what they can gain by their blood'.[3]

German culture was overwhelmingly oral. They kept no written records and maintained order through interpersonal relationships, oaths of allegiance and systems of mutual obligation. Ballads and poems that told the tales of past glories, honoured ancestors and celebrated gods were their sources of annals and history. Bards led soldiers into battle with raucous chants: the louder the din the better their chances. Conflicts were determined at regular assemblies where speakers were heard in order of precedence from the king or chief down. Displeasure was signified by a low murmur; approval by the thumping and clattering of arms.

The Germans had a form of alphabet known as runes – also derived from the Phoenician abjab – but it was reserved for magical purposes: the name rune comes from their word for secret.[4] Unlike the Romans, they didn't keep written administrative records or histories: knowledge passed directly from person to person.

Germanic spiritual beliefs were integrated into daily life rather than formalised into a recognised religion. They made great use of invisible signs and portents.[5] A favourite way to make decisions was to cut a twig from a fruit tree, divide it into pieces, and toss them at random onto a white sheet while invoking the gods. They would decide what to do based on where these 'chips' fell. At other times they would draw 'lots' or study the movements of birds or of specially kept sacred horses. They worshipped their gods and spirits in consecrated woods and groves, around wells and specific trees as well as in temples and shrines. They sacrificed animals and made offerings to Mother Earth so she would 'grow'. Amulets and charms were used for luck and protection against evil spirits.

The king of their gods was Woden, Tiw was the god of war, and Frigem the goddess of love. Thor was the god of thunder, Mani the god of the Moon and his sister, Sunna, the Sun. We hear the echoes of these deities' names in Monday, Tuesday, Wednesday, Thursday, Friday and Sunday. Saturday, for some reason, retains a root from the Roman god Saturn.

Tacitus' first-century portrait of the German people gives a preview of the next millennium of European history. In it we see a reverence for strength and valour in battle combined with an inclination to pursue the fight, the pre-eminence of personal and familial loyalty, and complex systems of interpersonal promises and obligations that were the precursors to the feudal system. Crucially, we see a people – soon to dominate all of Europe – for whom literacy was irrelevant and aesthetics mattered hardly at all.

For all that, the Germans were not in the least careless of what they saw. The visual world was extremely important to them, but it was the symbolic rather than the literal or representative that interested them. For the German tribes, the landscape was full of unseen entities that exerted power over their destinies and communicated with them through a myriad of visual signals. What they saw was *more* than what their eyes beheld; the seen world was teeming with hidden meanings and secret messages.

Western Europe, *c.* 600 CE

A couple of chaotic centuries after the fall of Rome, the Germanic migrations became settlements. Cities all but vanished, and international trade dried up. Schools closed and reading and writing fell into disuse. Warlords became aristocrats in their adopted lands and turned their attention to the acquisition of acreage, the key to wealth and power in an overwhelmingly agricultural barter economy. The European countryside became the patchwork of farms and manors populated by peasants and nobles that persisted for centuries. The old tribal system of loyalty and favour, fighting and feasting, plunder and reward pertained, dressed now in a more formal guise.

As the Germans gradually established themselves as noblemen around Europe, they also converted from their traditional beliefs to Christianity. Christianity had become the official religion of the Roman Empire in 380, seventy years after Emperor Constantine was converted by a divine vision of the Christian God. The Goths converted in the late 300s and the Frankish King Clovis was baptised a century later. The native Britons in Wales had been Christian since Roman times, and the Irish converted under St Patrick in the 400s. By the late sixth century only the Anglo-Saxon rulers of Britannia had yet to be Christianised.

From the Catholic Church's base in Rome, Pope Gregory looked to the former Roman province Britannia with interest. According to legend, Gregory had long been fascinated with the pale, blond-haired boys he saw for sale at a Roman slave market. When he enquired as to their origins he was told the boys were Angles from Northumbria. 'Not Angles,' he supposedly replied, 'but angels.' Now he was Pope he was determined to claim the souls of their kinsmen for Christ, and in 597 CE sent a team of missionaries to England do this work.

Pope Gregory advised his missionary party to:

By no means destroy the temples of the gods but rather the idols within those temples. Purify them with holy water, then place altars and relics of the saints in them. For, if those temples are well built, they should be converted from the worship of demons to the service of the true God. Thus, seeing that their places of worship are not destroyed, the people will banish error from their hearts and come to places familiar and dear to them in acknowledgement and worship of the true God.

Further, since it has been their custom to slaughter oxen in sacrifice, they should receive some solemnity in exchange. Let them therefore ... sacrifice and eat the animals not any more as an offering to the devil, but for the glory of God.[6]

So rather than destroying Anglo-Saxon shrines and altars, the missionaries converted them into Christian chapels. They adopted some of their festival days and invested them with Christian meaning, adding Christian names and symbols. The spring celebration of the Germanic goddess Eostre merged over time with the celebration of Christ's death and resurrection to become the Christian festival Easter. Christians adopted the Celtic festival of Samhain at the end of October, sometimes known as the Feast of the Dead, by moving All Saints' (Hallows') Day from May to November, making All Hallows' Eve a time to remember the dead. The midwinter pagan festival of Yuletide became Christmas.

Gregory's missionaries were employing the policy known as *Interpretatio Christiana* that had succeeded in converting pagans and their holy places, customs and practices to Christianity for centuries. Under this policy missionaries 'adapted' pagan customs to fit within Christian doctrine. Christianity had acquired some of its most sacred places – including the supposed sites of Christ's birth and crucifixion – from pagan sites that were 'baptised' under the *Interpretatio Christiana*.[7]

Missionaries found other ways to insinuate Christianity into local beliefs. In place of the numerous gods and ancestors the Germans venerated, Christianity offered the lives of the saints. Stories of saintly visions, revelations, persecution and martyrdom were presented with as much drama and heroism as those of the deeds of any ancient warrior. Just as many traditional gods had had specific responsibilities, saints were given patronage of particular activities. There were patron saints for each trade, for places, types of people, states of being, various illnesses and disabilities, emotions, and even pets. The cult of sainthood became and remained a key feature of medieval Christianity. Christians were forbidden from worshipping idols or using magic charms, but that didn't stop Catholic priests encouraging people to venerate the 'relics' associated with a particular saint, or from turning their burial places into sites of pilgrimage. Relics ranged from a piece of a saint's body – a bone or a finger perhaps – to a possession such as a robe or stick. It became common for saints to be dismembered after death and their body parts – heads, fingers, tongues, pieces of skin, bones – distributed, often encased in reliquary boxes highly decorated with precious jewels. Such reliquaries can be seen to this day in the treasure rooms of Catholic churches and cathedrals. The power of the relic was transmitted to the faithful through the eye: simply seeing a relic could trigger any number of miracles, from healing the sick to conceiving a child.[8]

Within a century of the Catholic missionaries arriving in Britannia all its kings and nobles were officially Christian. But it was several centuries before most ordinary people had access to a church or to regular sermons, even if they had been baptised. For hundreds of years traditional and Christian beliefs coexisted in haphazard blends of elements of the old faiths with the new, in rituals such as the Nine Herbs Charm.

Seeing in the Dark Ages

The Italian poet Petrarch coined the phrase 'Dark Ages' in the 1330s, referring to the period between the decline of Rome and a time he hoped for – not yet reached in his own day – when scholars would once again embrace classical learning. Earlier writers – and in fact Petrarch himself[9] – had used the metaphor in reference to pagan times before Christ's 'true light' appeared, but it was Petrarch's usage that caught the imagination of Renaissance scholars in the fourteenth and fifteenth centuries. Later Enlightenment scholars in the Age of Reason interpreted the epithet to refer to the dominance of irrational faith over reason during this period. In both cases the term was unapologetically pejorative.

As late nineteenth- and early twentieth-century historians learnt more about this long period, the term Dark Ages was parsed to make it less pejorative. 'Dark' now referred to informational rather than intellectual obscurity, brought about by the lack of historical records of the time. Today, the term has been all but banished from scholarly circles, though it is still used colloquially.

Whatever we call them, the centuries after the fall of Rome were unusually harsh and dangerous. We now know that there were extreme weather events in 535–36 CE, possibly caused by massive volcanic eruptions in the tropics, that led to decades of unusually cold weather, a steep decline in food production and worldwide famines.[10] In 542 an infectious disease broke out in an Egyptian port and quickly spread around the Mediterranean, along rivers and trade routes inland, and as far north as the British Isles, killing tens of millions of people. Over the next two centuries the disease appeared sporadically all over Europe before it eventually died out around 750. Descriptions of the hideous swellings and rapid death of its victims indicate this was the first epidemic of bubonic plague.[11] Smallpox was equally deadly and killed millions more.[12]

Add to these natural disasters the fact that the social and political structure relied on war for its survival, and it is unsurprising that the average life span was only two or three decades. Death was an ever-present possibility. Little wonder that Christianity, with its promise of resurrection and everlasting life in the next world, was so readily embraced. When life was so difficult and unpredictable, a preoccupation with the afterlife made perfect sense.

Seeing in the Dark Ages was both simple and extremely complex. Technology was almost non-existent; life was governed by the seasons and people had almost no control over their environment. There was no distinction between the natural and the supernatural, or between secular and spiritual life – all were part of the one reality, which included elves and dragons, divine intervention and satanic influences, and healing power in the shin bone of a saint, just as it included the miracle of a field of wheat growing up from a handful of seeds, the destruction of a village by a bolt of lightning, or the unexpected death of a healthy baby.

Seeking out links between events and trying to explain cause and effect are part of being human. In the harsh and uncertain world of the Dark Ages, when unexplained things happened every day, metaphysical explanations for life's random events must have seemed perfectly rational. Why shouldn't they be caused by a malicious spirit, or because someone did something wrong the day before? In the absence of any other information, it's as good an explanation as any.

Pre-Christian traditions were full of signs and auguries. Early Christianity's relationship with seeing was at least as complex. The Christian God is invisible, but various theologians grappled with idea of 'seeing' Him. St Paul wrote (in Corinthians I (13:12)) the famous passage: 'For now we see through a glass, darkly; but then face to face.' He was referring to seeing in a mirror, which at that time would give a dark and distorted image. It was only in the afterlife that Paul believed he would see God clearly, 'face to face'.

The early theologian St Augustine also wrestled with the nature of seeing and whether humans could ever 'see' God. He described states: seeing corporeally, with the eyes of the body; seeing spiritually, with the memory or imagination; and seeing with the eye of the mind or *mens*. The third was the highest form of seeing and the only way to experience divinity. Augustine wrote that when seeing with the eye of the mind, love and faith emanated from the mind's eye and reached out to God, then returned to the soul like a brief touch. In this life, mankind could only hope for a glimpse of God; it was only in the resurrected life hereafter that one could hope to gaze at God and truly see Him.

When it came to making images, early Christians were also divided. The first of the Ten Commandments states: 'Thou shalt not make unto thee any graven image, or any likeness of anything that is in heaven above, or that is in the earth beneath, or that is in the water under the earth.'

This was generally interpreted by pre-Christian Jews as referring to idols, but by the time of Jesus, Roman imagery had become extremely realistic. Paintings particularly almost gave the illusion of seeing the object depicted. Early Christian writers, including St Augustine, worried that such images could deceive viewers and lead to idolatry. On the other hand, many of the new converts to Christianity came from traditions that expected imagery as part of their devotions. The eventual solution was a change in artistic style from the illusory realism of classical art to a far more symbolic, abstract form of depiction. This became what we know as the Byzantine style: images with firm outlines, little depth and few details beyond what was necessary to convey the picture's message.

The Synod of 1025 confirmed the role of pictures as the 'literature of the laity', and images became a critical part of Catholic Christianity.[13] Citing Augustine, priests taught that when the eye's gaze fell on something it made a direct physical connection with the thing seen. Church walls were covered in painted scenes from the Bible, from the lives of the saints, of heaven and hell, and the Day of Judgement. They not only illustrated and reinforced biblical stories and lessons – worshippers also believed that, through the workings of the eye, images brought them into direct physical contact with Jesus and the saints.

Since at least the time when our Palaeolithic ancestors created art in caves, humans have had an imaginative connection with worlds beyond the visible present. The Greeks and Romans, with their pantheons of gods and rich mythologies, had strong religious traditions. They built magnificent temples to the glory of their deities, made offerings and celebrated festivals in their honour. But their achievements in the arts and engineering, philosophy, science and literature demonstrate that their religious beliefs did nothing to dim their interest in understanding the workings of the human and material worlds: the here and now. Life, and how it should be lived, was the topic of prime concern. What came next was incidental.

By contrast, medieval life was dominated by the prospect of the afterlife. There was no term for the supernatural, because the spiritual and mystical were as much part of nature as the corporeal. Death was feared, but not for the end of life. The fear was the risk of one's soul going to purgatory or, far worse, to hell itself.

Seeing was central to medieval life, but it wasn't seeing as we know it. It was seeing with the faith and imagination to make the unseen every bit as real as the seen. Intertwined with the material world were many things beyond view, but nevertheless entirely real. Invisible spirits were everywhere, some inhabiting living creatures, some within inanimate objects, others entirely ethereal. There were realms that existed beyond the familiar surroundings of home: the eternal kingdoms of the afterlife – heaven, hell and purgatory – but also lost lands with names like Atlantis and Brasil. These were supposedly real places, often shown on maps with topographical features such as mountains and lakes, and populated with fantastic beasts and wondrous phenomena.

Within the visible sphere, people were alert to signs and portents with significance beyond their optical appearance. Miracles and visions sent by God, His saints and angels, or else by darker and more ancient forces bent on deceit or destruction, were simply a matter of fact. And images made by men were also imbued with special power, as they could create an unseen yet physical connection between the viewer and that which the image portrayed.

More than any time before or since, this was the age of the invisible.

THROUGH A GLASS, CLEARLY:
SPECTACLES

For now we see through a glass, darkly; but then face to face: now I know in part; but then shall I know even as also I am known.

I Corinthians, 13:12, King James Bible

Northern Italy, Present Day

In a sleepy corner of a prosperous old city north of Venice, the Basilica San Niccolo towers over its squat post-war neighbours, a soaring Gothic survivor of the Good Friday air raid of 1944. Tucked away around the side of the enormous church is a former Dominican convent. A small donation at the front desk gives access through double doors to a peaceful cloister, on the far side of which a large oak door leads to the convent's former meeting room. Inside, forty portraits housed in cell-like frames form a frescoed cornice around the room's upper walls. Within each cell a painted metre-high figure in religious dress sits at a simple desk surrounded by books. There are cardinals in red hats, bishops in mitres, and monks with tonsured hair. A bored-looking nun has her chin in her hand. They are all reading, writing or copying manuscripts. A helpful twenty-first-century sign tells visitors that the fresco was painted in 1352 and portrays the most illustrious Dominicans of their day. The pictures are renowned, we learn, for their lively sense of humanity, striking realism, and freshness, and for the individuality of each subject.[1]

Up in the far corner sits a cardinal, hunched over his desk copying the manuscript sitting open on a shelf beside him. His right hand holds a pen, while his left holds a sheet of paper across the page on which he's writing as a ruler. His face is puckered. Is it concentration? Or is it to keep the armless, riveted eyeglasses perched on his nose?

This portrait of Cardinal Hugh of Saint-Cher, illustrious Dominican, is the earliest known depiction of spectacles.

All schoolchildren learn that a ray of light bends when it passes through two transparent substances of different densities. The phenomenon is called refraction, and we experience it every day in the way our eyes see: the cornea and lenses in our eyes refract light from the things we see onto the retina at the back of the eye.

Our eyes' lenses are naturally flexible, and muscles within the eye change their shape when we focus on objects at different distances. From middle age, however, and as a normal part of the ageing process, our lenses start to stiffen, making it harder for them to change shape and therefore to focus on objects that are up close. The first symptom of this change for most people is holding a book or a phone a bit further away to see it better. Reading small print becomes increasingly difficult over time.

A solution to this problem was discovered in the thirteenth century: clear convex lenses held close to the eye, by means of a frame, that refract the light coming into the eye and correct the refractive errors caused by the hardening lenses.

At least, that's what we know now. When spectacles were invented no one fully understood how they worked, or indeed how vision worked, and wouldn't for another three centuries. But their appearance wasn't random. In the two centuries leading up to the invention of spectacles, religious and societal changes placed a renewed emphasis on the visual, and a new requirement for visual clarity. Spectacles appeared at a time when, after a long period during which seeing was viewed with suspicion and trepidation, people wanted to look again at the world, and see it clearly.

Spectacles transformed the capacity of the thousands of people whose productive lives depended on close work. This included clerics and scholars, bankers and merchants, and seamstresses, goldsmiths and physicians. It can't be a coincidence that the leading artists of the time also sought a new clarity and realism in the way they depicted their subjects – a quest that would transform art and pave the way for the Renaissance.

Greece, 423 BCE

People have been using transparent substances to bend light for thousands of years. In the 423 BCE Greek comedy *The Clouds*, the roguish main character wondered whether he could escape his debts by sabotaging evidence:

> Strepsiades: Have you ever seen a beautiful, transparent stone at the druggist's, with which you may kindle fire?
> Socrates: You mean a crystal lens.

Strepsiades: That's right. Well, now if I placed myself with this stone in the sun and a long way off from the clerk, while he was writing out the conviction, I could make all the wax, upon which the words were written, melt.

Strepsiades was referring to a 'burning stone', a convex crystal lens that concentrated sunlight onto a single point and was commonly used as fire starter in ancient Greece.

A convex lens magnifies whatever is on the other side. It need not be solid. Four centuries after *The Clouds*, the Roman philosopher Seneca observed that everything appears larger when looked at through water, and suggested a glass globe filled with water could help with reading small writing.[2] His idea never caught on, however, and ageing Roman scholars continued to rely on slaves to read for them.

Archaeological findings provide tantalising suggestions that people may have used lenses as optical aids much earlier than this. Crystal lenses capable of providing a visible, magnified image have been found in a cave on Crete from perhaps 600 BCE, others from the site believed to be Troy several centuries before that, and in the Palace of Knossos from around 1400 BCE, though no one knows whether these lenses were used for magnification or decoration.[3]

Perhaps most intriguing of all is controversial author Robert Temple's report of microscopic Egyptian carvings on an ivory knife handle dated from 3300 BCE. The carvings purportedly depict human figures with heads 1 *millimetre* across (less than the width of a letter on this page) and recognisably braided hair. Temple asserts that since the carvings can't be seen with the naked eye, the Egyptian craftsmen who made them must have used magnification to execute the carvings, as must their patrons if they wanted to see the result.[4]

Despite these apparent cases of individuals using lenses for improving vision in antiquity, they did not spread beyond isolated cases and their use died out. They were not seen again for more than 1,000 years.

Europe, 600 CE

In the early Middle Ages the divine powers of God and the saints, as well as unseen beings and influences from pre-Christian mythologies, were ever-present, invisible realities. Seeing was an unreliable business in days when so much was invisible, and the Church explicitly taught people to distrust their physical senses. Even what was visible could be hiding several secret meanings. Was a magpie a bad omen, a messenger, or just a bird? Did a certain shaped cloud portend death or simply a rainstorm? The unseen was as real as anything in the visible world and was readily invoked to explain life's mysteries and vicissitudes. Meanwhile, the visual technologies that had flourished in the classical world – writing, painting, sculpture, architecture – had all but disappeared from everyday life in Western Europe.

There were important exceptions, however, where literacy, art and learning were preserved and fostered. The most important of these was the monasteries. In 529 CE, St Benedict of Nursia established the first monastery as a new type of institution within the Catholic Church. Removed from the harsh reality of the outside world, communities of religious men and women lived, worked and prayed together in monastic orders around Europe. Some were situated in remote and inaccessible areas, while others integrated themselves into their surrounding communities. Some were established with a specific religious purpose, while many were endowed by rich noblemen as penance for their warring and other sins.

Monasteries were governed by strict rules accounting for almost every minute of the day. In addition to the hours dedicated to prayer, there were specific times allocated for reading and manual labour. This dual requirement gave rise to a very particular aspect of monastic culture. Monks and nuns needed books, but they were difficult to come by and extremely expensive. The solution was that the most capable scribes spent their allotted labouring hours copying manuscripts word for word. This provided monasteries with books and at the same time fulfilled the monks' prescribed manual duties.

Copying manuscripts became a core feature of monastic life, and many monasteries built rooms called *scriptoria* dedicated to the purpose. The copying was the key – scribes were not allowed to make any adjustments or changes to the texts in front of them. It could be difficult and tedious work, and sometimes they had to copy texts in foreign languages or unfamiliar scripts such as Greek. Some monasteries had scholarly leanings but for many the content of the manuscripts they copied was of little interest: the priority was keeping potentially idle hands busy.

The great majority of the books copied in the monasteries were Bibles, theological treatises and other Christian texts. Occasionally, however, monastic libraries contained books preserved from classical Greek and Roman antiquity, most of which had been lost, destroyed or simply rotted away. Their pagan contents were ignored, but they were periodically copied and recopied, preserved in dusty libraries to be rediscovered many centuries later by humanist book hunters during the Renaissance.[5]

France, 800 CE

Not all medieval monarchs were solely preoccupied with war and plunder. On Christmas Day in the year 800, Pope Leo III proclaimed the Frankish king, Charles I, now popularly known as Charlemagne, Holy Roman Emperor. Charles had conquered most of Western Europe north of the Pyrenees.[6]

He was a brutal warlord but he was also deeply religious and keenly interested in learning. Charles despaired at the dismal state of literacy in his kingdom as both a religious failure – how could his subjects understand the scriptures and recite the liturgy if they couldn't read? – and because he realised he needed an effective system of administration for his empire. He recruited an energetic English cleric called Alcuin, reputedly the greatest scholar of his day, and set him the task of improving literacy in Latin and education generally throughout the kingdom. The Englishman set off establishing, improving and expanding schools, libraries and scriptoria in monasteries and cathedrals all around the Frankish empire.

One of the many challenges Alcuin faced was the array of handwritten script styles copyists used – many were illegible except to specialists. He introduced an Irish script that used simply rounded lower-case letters, and other Irish innovations including capital letters, question marks, and spaces between words. Classical Greek and Latin were written in continuous script – that is, without spaces – and this had continued into the Middle Ages. As one can imagine, this was very cumbersome to read. It required the reader to sound each syllable out loud to hear the words form themselves vocally. Scholars of medieval writing believe this didn't pose too much of a problem because reading at the time was an expert activity and was generally intended to be done out loud, to an audience. A professional reader typically prepared by deciphering the text beforehand. With spaces between words it became possible to read without having to decode each word out loud. Spaces removed the necessarily oral component of reading and made it possible for reading to be a purely visual activity. It also made it silent.[7]

Alcuin's new text was called Carolingian Miniscule. It made reading accessible beyond specialists and paved the way for it to become a private, individual activity. However, the Carolingian

Renaissance didn't last. Within three decades of Charlemagne's death his empire fragmented and local and regional lords once again took hold. Europe went back to being an overwhelmingly agricultural, feudal society. But the seeds had been sown for much more significant changes in the future. One outcome of Alcuin's project was a huge increase in the supply of written material in Europe in the 800s.[8] Literacy – in its new, more visual form – continued to spread slowly but steadily over the next few centuries.

Baghdad, *c.* 800 CE

Around the same time as Alcuin was building libraries and reforming literacy in France, 3,000 miles away in the new Islamic capital of Baghdad, the Abbasids of the new Islamic Empire were establishing libraries on an altogether more monumental scale. The Abbasids had conquered lands stretching west from present-day Pakistan around the southern Mediterranean as far as the Iberian Peninsula and were fabulously wealthy. They began collecting manuscripts from all over the known world. They gathered philosophical, medical and scientific treatises from Greece, mathematics from India, literature and philosophy from Persia. Scholars of all faiths and backgrounds were attracted to the great library they built in Baghdad, known as the House of Wisdom, where they could make a good living surrounded by learned colleagues.

Scholars at the House of Wisdom had a new, secret technology at their disposal. Instead of expensive and heavy parchment, they used an invention allegedly stolen from Chinese prisoners of war in 751.[9] The Chinese had been forming plant fibres into paper for more than 1,000 years. They had a rich reading culture and their imperial libraries boasted thousands of titles in tens of thousands of scrolls.[10]

The Arabs took the artisan techniques of Chinese paper-making and industrialised them as far as possible with the technology of the time. They prepared the pulp with a water-powered mill and used a manual trip mill to compress the pulp into paper.[11] To this they added a production-line-style copying process that allowed their libraries to expand to many thousands of volumes. For the first time in history it became economic to make and sell books on a grand scale; new trades in bookbinding and bookselling appeared, and writing and publishing flourished.

It what we now call the Translation Movement, scholars spent decades translating foreign texts into Arabic. When this was complete, they shifted their efforts to improving on the work of their foreign predecessors. Several turned their attention to the ancient body of work on seeing, known as Optics.

Some of the classical world's most famous thinkers had put their minds to the question of how we see. In Greece, Plato, Aristotle, Euclid, Galen and Ptolemy had all developed theories on how vision works. One school of thought held that vision was an invisible substance emitting from the eye to 'touch' the object seen. Another was that vision was a substance coming into the eye from the thing seen. A third held that emissions from the eye and the thing seen met somewhere in the space between them. None of these theories was even close to being right, partly because they all assumed there was a physical connection between the eye and the thing seen, and partly because vision is extremely complicated. It involves physics, physiology, geometry and psychology, and various aspects of brain science. To this day, scholars haven't mapped the entire visual system.

Another problem with the Greek theories was that each scholar approached the subject from their own area of expertise. Physicians such as Galen concerned themselves with the physiology of the eye and what could go wrong, with the aim of finding cures for eye disease. Philosophers such as Aristotle and Plato explored the nature of vision without offering any physiological explanation for how images are actually perceived. Meanwhile, the mathematicians, led by Euclid and Ptolemy, used geometry to attempt to explain how the shapes and sizes of the physical world could be translated into the mind through the medium of the eye, and addressed the related visual phenomena of reflection and refraction. None of the Greeks managed to bring these diverse aspects of vision together.*

Arabic scholars made several important optical discoveries in the ninth and tenth centuries. Al-Kindi (b. *c*.801) deduced that luminous objects emit light in all directions. Light is not, as had been thought, a substance analogous to a river but more like the ripples in a three dimensional pond. Ibn Sahl, writing in 984, described how lenses and curved mirrors bent light and developed a mathematical explanation of refraction, Snell's Law, that was only 'discovered' by Europeans in 1621.

Early in the eleventh century, the Arabic scholar Ibn al-Haytham, known in the West as Alhazan, turned his mind to

---

* Notwithstanding this, the Greeks had a highly sophisticated understanding of the practical aspects of vision, as is evident from their art and architecture. The 2,500-year-old Parthenon on the top of the Acropolis in Athens is a rectangular structure of eight by seventeen colossal columns. It appears to be a model of rectilinear perfection but it fact incorporates numerous visual tricks and imperfections that increase its appeal to the eye. The base isn't perfectly flat but gently domed; the corners are a foot or so below the middle; the columns all lean slightly inwards and are not of uniform diameter but bulge slightly in the middle and taper at the top. This makes them appear straighter than if they actually were perfectly straight (it's a phenomenon called entasis). The columns are not equally spaced either but are unequally spaced in a way that makes them appear more perfect. These refinements were not uncommon in Greek temples. Thanks to Julian Stevens, my eye surgeon, for pointing that out.

optics. Alhazan was a prodigious polymath who wrote as many as 200 works on all aspects of science, among which were more than a dozen on subjects related to light and vision.[12] Unusually, he relied on systematic observation rather than received wisdom and reasoning, and conducted experiments to test his theories,[13] and for this reason he has been called the first scientist and true father of the Scientific Method.[14]

In one experiment, Alhazan set up a darkened space with a flat wall on one side and an opening on the other. He lit candles outside the space and observed their effects on the wall inside. When the opening was large, the image on the wall was the same shape as the window. This confirmed al-Kindi's earlier theory that light emits in all directions. However, as Alhazan reduced the size of the opening, at a certain point the image on the inside wall changed. Instead of a window-shaped pool of light there were individual candle flames visible on the wall. By covering each candle in turn he discovered that the flames' images were inverted and reversed left–right, leading him to conclude that light rays from the candle flame travelled in a straight line from the flame, through the hole and onto the wall behind, crossing over each other at the hole. It is not clear whether Alhazan did further experiments himself, but a later Arabic optics scholar demonstrated that an accurate, colour image of any brightly lit scene or object could be projected through a small hole onto a flat surface in a darkened space.[15]

This phenomenon, later called a *camera obscura* from the Latin for darkened room, had been observed since antiquity – Aristotle commented on crescent-shaped images appearing through broad-leafed trees during a solar eclipse – but Alhazan was the first to explain it. His experiments provided vital insights into the nature of light and, as we now know, vision, as it explains the way the eye forms images on the retina.

Alhazan stopped short of directly likening the *camera obscura* to the eye but he did initiate the comparison when he wrote:

> the form's [image's] arrival in the common nerve [the eye's seat of vision] is like the light's arrival from windows or apertures, through which light enters, at the bodies facing those windows or apertures.[16]

Alhazan concluded that light travels in straight lines or rays, which is why shadows form when there is an obstruction between the light source and a surface. This characteristic of light – called rectilinear propagation – may seem obvious but consider that most moving things travel either in an elliptical curve as they are pulled down by gravity – like a ball or an arrow – or in waves that go around corners and obstructions – like water or sound. Alhazan proposed that vision occurs when light rays travel from the object seen to the surface of the eye and are refracted there, before coming together in the centre of the eye to form an image. The image the eye receives consists of only light and colour, he said. All the other aspects of vision – distance, size, beauty and so on – must be understood separately, with the mind.

With these insights Alhazan quashed all previous theories and synthesised the three branches of classical knowledge – the physiological structure of the eye, the nature of vision and light, and the geometry of perception – into a single theory for the first time.[17] His theory wasn't entirely accurate; he didn't fully appreciate the similarity between the *camera obscura* and the eye, for example, and he put the eye's seat of vision in the lens rather than the retina, but it was a remarkable advance.

Five hundred years later Leonardo da Vinci made the first direct comparison between the *camera obscura* and the eye, and another century after that Johannes Kepler completed the analogy when he correctly explained the roles the cornea, lens and retina played in seeing for the first time. Sir Isaac Newton's *Opticks* took another 100 years to appear.

For the centuries in between, Alhazan's *Book of Optics* was the closest mankind had come to understanding light and vision. Its consequences for the history of seeing were monumental.

Paris, 1140 CE

Between the years 900 and 1300, an unusually warm climate, relatively stable politics and improved agricultural methods increased prosperity in Western Europe. Trade resumed, populations grew, and towns and cities formed around a mercantile, trading economy.

By the twelfth century, society had expanded from the threefold medieval structure of nobles, clergy and peasants to include middle-ranking urban folk such as artisans, merchants, teachers, bankers and lawyers. Many of these people used basic reading, writing and arithmetic in their daily lives, either as an intrinsic part of their work or for keeping accounts and recording agreements. The new middle classes formed organisations such as town councils, trade guilds and local associations independent of the Church and the nobility, to regulate their industries and manage their crowded living spaces. Literacy and numeracy moved outside the rarefied atmosphere of the monasteries and onto the streets, and in many cases regional spoken languages – the vulgate – displaced Latin as the language of written communication.

Living at close quarters, the new bourgeoisie became increasingly conscious of how they were 'seen' by their fellow townsfolk. Conspicuous consumption increased, as did conspicuous piety; religion remained a central part of people's lives at all levels of society and fear of eternal punishment in the afterlife remained a preoccupation. Lavish contributions to the Church displayed both wealth and piety, making the Church a major beneficiary of the rising prosperity. In the growing cities, a huge programme of church building and enlargement ensued.

The Romanesque (called Norman in Britain) architectural style that had predominated across Europe throughout the early Middle Ages was emphatically solid, and featured thick, rounded walls and small, narrow, arched windows. In the twelfth century, technical advances in the flourishing Venetian glass industry dramatically increased the availability of glass, while a radical architectural innovation made it feasible to build much larger windows than ever before.[18]

In 1140 the French Abbot Suger began expanding Saint-Denis church just outside Paris. His architects designed flying buttresses and rib vaults that allowed walls to be thinner and higher with enormous windows filled with clear and stained glass. These were intended to 'brighten the minds, so that they may travel, through the true lights, to the true light where Christ is the true door'.[19] Contrary to early medieval theologians, Abbot Suger believed passionately in the power of seeing beautiful things to uplift the soul and take it to 'some strange region of the universe which neither exists entirely in the slime of the earth nor entirely in the purity of Heaven'.[20] His church, Saint-Denis, was the first example of Gothic architecture and inspired similarly towering, light-filled churches – also filled with ornate treasures bought by generous, guilty patrons – all around Europe.

As light flooded into the new Gothic churches, new light was also being shed on religious practice. The Franciscan and Dominican orders formed in the early 1200s eschewed monastic life and travelled around preaching to the masses in the new urban centres using vernacular languages instead of Latin. Their approach was emotional and direct, in contrast to the remoteness of the mainstream Church. They exhorted their followers to visualise the stories of the Bible and the saints; to imagine what it must have looked and felt like to find a baby in the bulrushes, to face down lions or to flee to Egypt. St Francis himself was a lover of plays and theatre, and he set the ultimate example on Christmas Eve, 1223, when he brought together a manger, straw, an ox and a donkey to create the first ever Nativity scene in a cave in Grecci so his followers could see 'with bodily eyes the inconveniences of His birth'.[21]

Religious emphasis shifted from the esoteric and mystical to a more straightforward narrative to which people could relate. Priests stressed the humanity of the principal characters in the Christian story and encouraged the faithful to express their empathy with them. Sermons and depictions presented Jesus and Mary as real people with human characteristics rather than remote figures of divine majesty. Mary's role as a mother came to the fore: nursing a suckling baby, bringing up a precocious child and – eventually and inevitably – losing her beloved son, as so many other mothers of the time would have done. The iconic image of the *pietà* – the tragic Mary cradling her dead adult son – first appeared around this time and became a recurring theme in art.[22] Mary's own life story was also a popular theme and was a common subject of fresco paintings decorating the new Gothic churches. Jesus Christ was increasingly portrayed as a real man who suffered pain and humiliation at his betrayal and crucifixion, as any man might, rather than as the triumphant King of Heaven sitting on a throne next to God. The mutual tragedy of mother and son was real, familiar and personal, evoking a directly emotional response rather than the generalised awe of the past. Religious experience became less abstract and much more a part of the seen, immediate world of the here and now.

Meanwhile, outside the Church in the brightest young minds of the day, a different type of enlightenment was taking place.

The Crusades of the eleventh and twelfth centuries brought Europeans into contact with the Islamic world. Tales soon circulated of the great Arab schools and libraries of Baghdad, Alexandria and Cordova, their access to the great works of antiquity, and the Arabs' great advances in scholarship. Curious young Europeans came together to seek out knowledge and teachers, and a new type of institution appeared, first in Bologna (1088), then Paris (1150) and Oxford (1167). Universities, as they became known, were free to explore subjects beyond the concerns of the Church. All they needed was material.

Luckily for them, material was not long coming. As European forces conquered Islamic cultural centres in central Spain and Sicily they brought some of the great libraries of Arabic texts under Christian control. They also gained access to the secrets of making paper. In a reversal of the Translation Movement three centuries earlier, Arabic texts were translated into Latin, releasing first a trickle then a torrent of scholarship from the Islamic world into Europe. European scholars were reintroduced to classical works that had been lost since the decline of Rome, to which had been added the knowledge the Arabs had accumulated from India and China and the advances their own scholars had made.

The most significant of the classical rediscoveries was the work of Aristotle. For seven centuries Christian thinking had been based on the teachings of fourth-century theologian St Augustine, who was in turn heavily influenced by Plato. Augustine discouraged reliance on the senses, believing that only the mind and soul could determine truth or access knowledge. The senses, he wrote, were easily deceived and could lead the body into sin.[23]

Aristotle's belief system was the opposite. He taught that the senses, especially seeing, were essential to understanding the world. Scholars adopted his position in droves, and the emphasis of scholarly discourse shifted from the internal to the external, from the soul to the senses, from the general, idealised form to the particular, from the group to the individual, and from the spiritual to the physical. Aristotle's teachings prompted a profound change in outlook among twelfth- and thirteenth-century scholars. They opened their minds and eyes to looking at, and seeing, the world around them.

Alhazan's *Book of Optics* was translated into Latin by an unknown scholar around 1200. It prompted a flurry of interest in the subject. Eminent European scholars including Roger Bacon, John Peckham and Witelo wrote treatises on optics, drawing heavily on Alhazan. It was a subject of keen interest at the Papal Court.[24] Bacon remarked that lenses 'will prove to be a most useful instrument for old people and all those having weak eyes, for they can see in this manner the small letters'.[25] He was probably referring to magnification of letters rather than spectacles, but the thought was prescient.

Scholarly discourse on optics stalled after about 1280, but the resurgence of interest in seeing had seeped out beyond scholarly walls. Practitioners working in the visual crafts – artisans and artists – began their own investigations into light and vision, with historic results.

Pisa, *c*. 1278

The first optical device emerged in the tenth century, when reports appear of monks and nuns using 'reading stones' to aid reading. These were essentially magnifying glasses: spherical or plano-convex (curved on top and flat at the base) rock crystal lenses that they placed over letters to make them appear larger.

In 1278 or thereabouts, just as the scholarly opticians were presenting their versions of Alhazan's insights to the Papal Court, an anonymous artisan in Pisa came up with the idea of putting two convex lenses into frames and joining them with a rivet so they could rest on the nose. This invention – eyeglasses, or spectacles – literally transformed seeing.

Eyeglasses work quite differently from a magnifying glass, which simply enlarges what is being looked at. Eyeglasses correct faulty vision by repositioning the eye's point of focus so that the visual image lands correctly on the retina. For the long-sighted, glasses with convex lenses bring the focal point back to hit the retina properly. For myopia or short sightedness, concave lenses move their focal point forward. That is, eyeglasses don't change *what* the viewer sees, like a reading stone, they change *how* they see.

Remarkably, this revolutionary new use of lens technology was discovered by serendipity, since no one understood the visual process well enough to have known how to intervene in it directly. What's more, the person who invented eyeglasses wasn't an inventor – there wasn't even such a word in Italian at the time.[26] Nor were they a scholar or a doctor. All we know is that they were a craftsman, perhaps a goldsmith, or an engraver or toolmaker.

How did they arrive at the idea of eyeglasses? According to Dominican convent records, the person who invented eyeglasses didn't want to share the secret of their invention with anyone.[27] Was this reluctance in order to retain a monopoly on the device to their commercial advantage, or was it perhaps simple lack of interest? No one knows. Fortunately for the rest of the world, a handy Dominican friar from a nearby convent crossed paths with the reticent 'inventor' and his eyeglasses, and promptly copied them. Alexander della Spina carried on making spectacles for the rest of his life and, according to his obituary, 'showed everyone else how to do so with an open and joyful heart'.[28] The inventor must have been delighted.

Once discovered, spectacle making spread quickly throughout Italy, probably via the peripatetic Dominicans. In 1289 the

Florentine Sandro di Popozo wrote, 'I am so debilitated by age that without the glasses known as spectacles, I would no longer be able to read and write.'[29]

In 1284 the powerful Glassmakers Guild of Venice – the glass-making capital of the world – made reference in their records to 'magnifying lenses' and 'discs for the eye'. Their 1300 regulations mentioned eyeglasses – *ochiali* – and in 1301 they added a clause allowing anyone to manufacture or sell 'eyeglasses for reading' if they fulfilled certain conditions.[30] By 1316 a basic pair of spectacles could be bought in Venice for about the cost of a pair of work shoes, making them affordable for most working people. Spectacles had certainly reached England by 1326, when they were mentioned in the inventory of a deceased bishop. By 1391 they were being imported into England by the thousand: almost 4,000 pairs came into London in May that year alone.[31]

Early spectacles all had convex glass lenses that correct for long sight and presbyopia. They were similar to the reading glasses we can now buy off the shelf for just a few pounds, but imagine how precious they must have been, and what a revelation. Deciphering handwritten books and documents must have been challenging for anyone, let alone those with declining eyesight. But it wasn't just books. Everything people used and wore was made by hand, and the people who did the making – weavers, seamstresses, tailors, cobblers, goldsmiths, blacksmiths, apothecaries, to name a few – needed to see what they were doing. Then recall that light levels inside were much lower than we enjoy now, as windows were smaller and there was only firelight or a candle after dark.

Even those who haven't experienced presbyopia will appreciate that eyeglasses must have been transformative – as they still are – for anyone in their middle years who had to do close work. Before the invention of spectacles, craftsmen, clerics and scholars were routinely thwarted by age just as they reached mastery of their profession. Eyeglasses could extend their productive lives by years and prevent this natural attrition of expertise. One can readily appreciate what a boost to productivity this must have been.

While we don't know the precise circumstances in which spectacles came about, we do know they appeared at a time when seeing was in the ascendance after centuries in the wilderness. Did a learned Italian who had read or heard of Bacon's work, or perhaps the work of Alhazan himself – one of the scholarly Dominican friars perhaps? – find himself discussing these ideas with a surly middle-aged craftsman who was struggling with his own declining eyesight? Did that craftsman take it upon himself to fashion a practical device to help him see while he worked so he could eke out a few more productive years? It wouldn't really matter that he didn't properly understand vision. Or perhaps the craftsman had no inkling of the circulating theories and was simply trying to fashion a handsfree reading stone? Trial and error, experimenting with lenses on his own eyesight – glasses are, after all, fairly simple objects, despite the complex theory behind them.

We will probably never know for sure. But we can say with certainty that glasses have transformed lives ever since. It was nearly two centuries before concave spectacles that corrected short sightedness, or myopia, appeared in the 1450s. The delay may have been because concave lenses are more difficult to grind. Or perhaps it was a reflection of the fact that no one actually understood how spectacles worked, so didn't think to try concave lenses. In any case, concave spectacles allowed people with the hereditary condition of myopia to enter into productive life. Myopia was believed to be quite rare in the Middle Ages but it has become an increasingly common condition. Researchers don't completely agree on the reasons for it, though one accepted cause is spending less time outside. Today myopia affects 23 per cent of the world's population, and a recent forecast estimated that this will increase to 50 per cent by 2050.[32] A world without spectacles is simply unthinkable.

Padua, 1302

Heightened interest in observation and the shift in religious emphasis from the abstract to the tangible began to make an impact on art in the twelfth century. In sculptures adorning the vast new cathedrals, artists began to create more lifelike depictions, adopting classical techniques to show the forms of the body and expressing emotion in posture and facial features.[33] Artists started using more subtle modelling (adding highlights and lowlights to suggest contours) in painting and including observed details to illustrate familiar stories. This was a move away from the standardised images their colleagues had copied for centuries.

In 1302, just twenty years after spectacles were invented, a wealthy financier named Enrico Scrovegni commissioned a young Florentine painter to decorate the chapel he had built adjacent to his luxurious palace in the university town of Padua. Giotto di Bondone was in his 30s and already one of the most famous artists of his day, having worked for popes and princes in Rome, Florence and elsewhere. According to legend, the master artist Cimabue discovered Giotto as a shepherd boy drawing his sheep on a rock surface. The sketch was so lifelike, Cimabue invited Giotto to become his pupil. One day, while his master was out, Giotto painted a fly on the nose of the painting Cimabue was working on. It was so realistic that when Cimabue returned he tried several times to brush the insect off.

The subject matter for the Scrovegni Chapel was a typical of its time: a fresco cycle depicting the lives of the Virgin Mary and Jesus Christ, and a monumental Day of Judgement painting on the western wall surrounding the main door (to remind congregants to consider what they had in store for them on their way out). The subjects were traditional but Giotto's execution was revolutionary, and the project illustrates several of the changes in worldview under way.

Firstly, Giotto's patron, Scrovegni, was from the new classes – a lay person from the commercial world outside the nobility and clergy. He was immensely wealthy but also, as a banker, inherently suspect of the sin of usury (lending money at extortionate rates of interest). The chapel was both a vivid display of his wealth and a recompense to God to expiate his sins. As was becoming common at the time, Scrovegni himself was depicted in the Day of Judgement scene – situated among the blessed naturally, not the damned – presenting a doll-house-sized version of the Chapel to the Madonna.

Secondly, Giotto's treatment of his subjects diverged from the Byzantine tradition of previous centuries in both form and technique. Giotto structured his series as a narrative, each scene contained in a painted frame like a graphic novel. He used realistic compositions, arranging the figures according to the logic of the narrative rather than their religious importance. In the Nativity scene, for example, Mary and the baby Jesus are to one side of the picture rather than in the middle, as would usually be expected for the scene's most important protagonists. Mary is presented almost in profile, in a radical departure from the front-facing Byzantine tradition. Figures elsewhere have their backs to the viewer where this makes sense to the scene. The subjects communicate with one another, looking into each other's eyes rather than staring piously ahead. Their faces, gestures and postures express intense emotion, including streaking tears, carefully placed hands, and hunched shoulders, all designed to stimulate the viewers' emotions. Backgrounds – buildings, trees, distant hills and rivers – are depicted in detail to situate the characters in a real time and place, rather than having them float enigmatically in space. All these details seem natural to us now but were extraordinary at the time.

Most radical of all, perhaps, was Giotto's technique. Giotto attempted to present his subjects in three dimensions, illustrating space, mass and volume on a flat surface using light, shade and geometry for the first time in 1,000 years.[34] Rather than have his scenes illuminated evenly as was usual at the time, Giotto chose a notional light source – in this case the western window of the chapel – and painted every picture as if it were lit from there. This knack was taken much further by later Renaissance artists, but it all started with Giotto. He also painted his scenes from a notionally fixed viewpoint, employing an early version of linear perspective more than a century before Brunelleschi 'discovered' it in Florence.[35]

Giotto brought a Franciscan worldview – with its love of nature grounded in visible, earthly reality – to religious art, and used techniques that made his two-dimensional painted images more like reality than anything seen since Roman times. It is impossible not to speculate, as some modern scholars have, that there was a connection between these new techniques, the treatises on optics that were circulating at the time, and the introduction of spectacles to the glass trade in neighbouring Venice.[36] Padua was a university town. Lectures on colour, light and the geometry of vision were being given there at the time Giotto was working on the chapel.[37] Giotto may have attended some of these lectures or, perhaps more likely, discussed the concepts with professors and students in the town's tavernas. He may also have discussed them with his close friend, the poet Dante, who had studied science and medicine and was fascinated by optics. Dante referred to reflection, refraction and presbyopia in *The Divine Comedy*.[38] The coincidence in the timing of Giotto's visual breakthroughs in painting with other important optical developments seems too close to be random.

Bruges, 1434

As Italian painters were experimenting with Brunelleschi's ideas of linear perspective, in the Netherlands, a Flemish artist called Jan Van Eyck was painting a portrait of the Italian cloth merchant Giovanni di Nicolao Arnolfini and his wife. The painting depicts a sumptuously dressed couple standing facing one another in a well-furnished room with a large window to one side. A metal chandelier hangs from the ceiling and there is a convex mirror on the rear wall in which one can see the reflection of the couple and two other figures in a doorway, one of whom is presumably the artist. It is now in the National Gallery in London.

The painting looked completely different from anything being painted in Italy at the time. It was extraordinarily realistic, in terms of its apparently accurate linear perspective, its detailed rendering of the couple's clothing and other surfaces, and the depiction of the couple's faces. Until now painted faces had not been particularly differentiated. It had been normal since Giotto to convey emotion via facial expressions, but even so each artist tended towards a generic facial shape and features with a definite stylistic, pictorial quality. The Arnolfinis' faces, by contrast, looked like real people rather than paintings. The same was true of their clothing, which suddenly not only had depth, but also elaborate patterns whose details were accurately captured in the twists and folds of the gorgeous fabrics. Other Flemish paintings of the time have a similar photographic realism.

Until recently modern art historians put this change down to Van Eyck's use of newly invented oil paints, unknown in Italy at the time. Oil paints have richer colours and can be applied more slowly and with more layers than is possible with frescoes – which have to be made on wet, fresh plaster, hence the name. Oil allows for more subtle changes in tone and more time to add detail. It was assumed that Van Eyck, who had a background in painting miniatures so was adept at intricate details, had employed the new paint medium to virtuoso effect.[39]

But in 2001 the renowned British artist David Hockney and optical scientist Charles Falco published their theory that Van Eyck was using optical devices to achieve his extraordinary results. They contend that Van Eyck used an optical projection device, probably a *camera obscura*, to project images onto a surface from which he could quickly and accurately trace or plot the subject's key outlines and features, or even paint directly over the projected image. Van Eyck was a learned and well-connected man, more than capable of accessing the circulating optics treatises, and they believe he took the principles of the *camera obscura* described by Alhazan and Bacon and made his own version, possibly improving the quality of the image by using a concave mirror. In this way, Hockney and Falco believe Van Eyck and other Flemish artists created hand-made 'photographs' without film.[40]

Within a couple of decades, the Hockney–Falco thesis continues, the Flemish artists' secret knowledge spread to Italy. Lenses replaced mirrors in the *camera obscura* late in the sixteenth century, when their quality improved sufficiently to provide a sharper image, and intensely lifelike painting became the norm until the nineteenth century, when the invention of chemical photography prompted artists to seek new forms of visual expression.

Hockney is careful to point out that his theory doesn't detract from the mastery of the artists – mirrors don't make marks, he said, artists do – but nevertheless the thesis attracted significant controversy and criticism, and prompted a debate that is not yet resolved.[41]

What of the evidence? It is well established that artists used mirrors from at least the time of Brunelleschi's 1420 experiment in Florence. In 1434 Alberti wrote, 'it is remarkable how every defect in a picture appears more unsightly in a mirror. So things that are taken from nature should be emended with the advice of a mirror.' Leonardo da Vinci gave similar advice, saying, 'when you are painting you ought to have by you a flat mirror in which you should often look at your work. The work will appear to you in reverse and will seem to be by the hand of another master and thereby you will better judge its faults.' He went on to explain that a mirror image can help an artist 'see' the trickier aspects of perspective, such as foreshortening an outstretched object.

In terms of using projected images, scholars have claimed for 100 years that Vermeer used a *camera obscura* to render his lifelike images in the seventeenth century.[42]

But the Hockney–Falco thesis contends optics were used more than 200 years before this, and much more widely. There is no written evidence for their theory, so the pair pointed to the paintings themselves to make their case. Hockney drew on his artistic expertise to point out where the new 'photographic' style differs most significantly from the traditional 'geometric' approach, and Falco analysed images to identify details that would be consistent with the use of optical projections. In the Arnolfini portrait, for example, Falco used a computer model to adjust the six arms of the chandelier as depicted for perspective. He discovered that, once adjusted, they were the same size within 1.5 per cent for width and 5 per cent for length. In a complicated object, he concluded, this perspective presentation was just too perfect to have been achieved by eyeballing alone.[43] In other paintings he identified slight shifts in perspective, and pointed out these were consistent with moving the projection or changing focus.

Hockney also noted the predominance of paintings, especially portraits, framed by a window ledge, with dark backgrounds and strong directional light – and pointed out these are all conditions consistent with a *camera obscura* set-up. Critics of his theory claim, among other things, that the quality of mirrors and lenses wasn't good enough at the time to project images, and challenge Hockney's interpretation of perspective in the images he used as evidence.

Whether or not Van Eyck's and others' paintings were made using optics, one can imagine the sensation they must have created for fifteenth-century viewers. Seeing images almost as real as if they were viewed through an open window must have been both uncanny and awe-inspiring. The religious works made for churches personified Jesus and the saints and brought the word of God alive like nothing before them. And, no wonder, they prompted a huge rise in portraiture that carried on throughout the Renaissance. Imagine how a lifelike portrait, clearer than a mirror image of the time, must have appealed to the vanity of those who could afford one. Not only did it depict one's very own features but could also show rich fabrics lined with fur, glistening armour and twinkling jewels and other objects of desire and prestige, and luxurious interiors. All the wealth and ostentation of the time could be captured conveniently in a single frame and put on display forever more.

And enjoyed well into old age, with the aid of spectacles.

Treviso, 1352

On the plain between the coastal archipelago of Venice and the foothills of the Dolomite Mountains sits the prosperous Italian town of Treviso. In 1340, after years of battles between warring local families, it became part of the Most Serene Republic of Venice. Along with welcome peace, membership of the Republic brought the city great wealth as it piggybacked on Venice's success as a marine superpower and the glass-making capital of the world.

Alongside the city's commercial boom, its churches were also growing. The city boasted a large Romanesque cathedral built on the site of a Roman temple and several newer churches and convents built in the Gothic style. There was no shortage of wealthy donors ready to commission decorations for the new buildings, creating a vibrant market for artists and artisans. The most talented artist working in the city was a young painter from Modena called Tommaso Barisini. In 1350 he completed a painting of Mary and Seven Saints for Treviso's Franciscan church. The painting's unusual level of realism and intimacy prompted comparisons with the work of the late Giotto, painter of the famous fresco series in nearby Padua fifty years earlier.

Not to be outdone by the Franciscans, the Treviso Dominicans commissioned Tommaso to decorate their chapter house in preparation for an upcoming gathering of their order. Tommaso's brief was to celebrate the Dominicans' scholarship and intellectual achievement. The artist decided to approach his brief literally. He conceived an imaginary scriptorium with all the most famous Dominicans sitting in their cells studying or copying manuscripts. Within the rigid template he had designed for himself, Tommaso determined to depict each individual uniquely. This was a challenge as many of his subjects were dead or distant and he had no way of knowing what they actually looked like. With an artist's ingenuity and imagination, he assigned each grandee their own gesture or posture or facial expression, or particular tools of their cerebral trade. Most had pens and inkpots, but there were also scissors, rulers, a magnifying glass, and reading stones.

One of Tommaso's subjects was Hugh of Saint-Cher, a French priest from the previous century who was the first Dominican to be made a cardinal. Among many other achievements, Hugh had prepared the most comprehensive index of words used in the Bible ever seen. Hugh must have displayed a prodigious attention to detail, and it was perhaps to emphasise that characteristic that Tommaso decided to depict Hugh wearing spectacles, even though he had died twenty years before they were invented.

OBSERVING

THE OPTICAL TOOLS
THAT MADE THE
MODERN WORLD

GUNPOWDER FOR THE MIND: THE PRINTING PRESS

What gunpowder did for war, the printing press
has done for the mind.

Wendell Phillips (1811–84)

Constantinople, 1453

On Tuesday, 29 May 1453, the 21-year-old Ottoman Sultan Mehmed II rode his white horse into Constantinople's vast and ancient domed basilica, the Hagia Sophia, and claimed it for Islam. It was the conclusion of a forty-day siege during which the precocious warrior had bombarded the city's mighty walls with the largest cannons the world had ever seen. The young Turkish ruler blockaded the surrounding seas to thwart any maritime relief arriving and, when faced with a huge iron chain blocking entrance to the Golden Horn, had his troops drag their ships up and over the steep hill of Galata, past the defensive chain and into the city's harbour. For more than 1,000 years Constantinople – also known by turn as Byzantium, the New Rome, and simply the City[1] – had been the Eastern capital of Christianity, the link between Europe and Asia and the main Western access point to the lucrative Silk Road trade route. The fall of Constantinople sounded the death knell for the Byzantine Empire, the last outpost of the Roman Empire, and the beginning of half a millennium of Ottoman rule in the East.

Later that same year, at the opposite end of the European continent, the city of Bordeaux – ruled for the last 300 years by the English kings of the House of Plantagenet – surrendered to the French following defeat at the Battle of Castillon. Like the Turks, French forces triumphed by deploying new gunpowder-based artillery, in their case using guns to shoot English longbowmen to pieces. The event was the last in a series of Anglo-French battles known as the Hundred Years War. England became an island nation, offshore and separate from Europe, and soon fell into another long-running dispute, the internal War of the Roses.

Meanwhile, in the south-western corner of the continent, Portuguese ships were bringing the first African slaves to Europe, along with considerable quantities of gold. Prince Henry the Navigator's explorers had found a sea route to Cape Verde off the West African coast. This was an important staging post in their quest to circumvent the Arab-controlled land routes across the Sahara and find a sea route to India. The Age of Discovery was stirring on the Iberian Peninsula.

As these momentous events were taking place at Europe's frontiers, deep in the heart of the continent an obscure goldsmith was working on an invention that would eventually change the world more profoundly than any war, battle or colonial conquest.

Mainz, Germany, 1453

After years of experiments with equipment and materials, failed business ventures and legal entanglements, by 1453 Johannes Gutenberg felt sure that the new method he had devised for producing books would finally make his fortune. Gutenberg had developed a machine that could copy books by pressing or 'printing' lines of text formed from individual letter 'types' onto a page. Until now every book, pamphlet, document, contract and letter ever produced had been written by hand, page by page, copy by copy. For millennia, scribes had sat hunched over cramped desks, mixing pots of ink, sharpening pens and spending miserable days forming line after laborious line, with just a few sheets to show for it by day's end and with endless possibility for error. By contrast, a single Gutenberg machine could print more than fifty sheets a day, each one identical to the last.

From his premises in Mainz, Germany, at the confluence of the mighty Rhine and the more sedate Main rivers, Gutenberg was working on printing an edition of complete Latin Bibles. It was a highly ambitious project. Each copy comprised nearly 700 sheets, and more than 1,200 pages, bound into two volumes. To fund the enterprise, Gutenberg borrowed 1,550 guilders[2] – a significant sum – from local financier Johann Fust. He set up a large workshop with six presses running in parallel and employed teams of men to operate them. He planned to produce 180 copies over two years.[3]

As Gutenberg's project was nearing completion, disaster struck. His backer, Fust, brought an action against him for repayment of his loan, plus interest. It is not clear why Fust did this when to all intents and purposes the project was proceeding according to plan. Some speculate it had been Fust's intention all along. On the other hand, Gutenberg had a history of antagonising business partners, so perhaps he had provoked Fust, too. Whatever its merits or motivation, Fust's action succeeded and he was awarded control of most of Gutenberg's printing equipment, and all the completed Bibles. Fust and one of the workshop's employees, Peter Schoeffer – who married Fust's daughter Christina soon after – went into business together and went on to establish the world's first printing empire.

Meanwhile, Gutenberg had to start from scratch. He found another backer and continued printing on a small scale but he seems to have been a fairly hapless businessman, and never achieved fame or fortune in his lifetime. Gutenberg died in obscurity in 1468, but he did have the last laugh. His achievements were eventually recognised and the world now associates the name Gutenberg with the invention of the printing press. Fust, on the other hand, was later confused with the devilish Doctor Faustus,[4] while he and Schoeffer remain under eternal suspicion of conspiring all along to steal Gutenberg's invention from under him.

Gutenberg's achievement was to mechanise an artisan craft, even though early printing workshops would look very artisanal to us. The press itself was a heavy wooden contraption about the size of an average bookcase, with an H-shaped vertical frame straddling a horizontal bed. Each press required three workers. The compositor laid out the text, letter by letter and line by line, with individual metal letter types, using a compositing stick not unlike a Scrabble letter rack. The types were cast in mirror image from metal matrices, made from carved wooden punches, and stored carefully in cases. Capital letters were stored in the upper case, and ordinary letters in the lower case. As the compositor completed each block of text, he transferred it into a small tray called a galley. When a page was complete, a proof was taken, and corrections made before the text was locked into a frame on the press. Next the inker, or 'printer's devil', used a stuffed leather ball with a wooden handle called a dabber to apply an oil-based ink to the raised letters. The third worker, the pressman, laid a sheet of damp paper over the text, folded a padded wooden plate over it, and screwed the plate down evenly onto the inked letters with a screw mechanism adapted from a wine or olive press. The printed impression was now the right way around. Two pages were printed on either side of a sheet and folded to form four page-sides.[5]

Printing wasn't new. People had used stamps and seals since Babylonian times, and there was a centuries-old tradition of woodblock printing in Asia. Various artisans in China and Korea experimented with moveable type from the eleventh century, but it never caught on, perhaps because of the large number of characters in those scripts. In Europe, people had been using woodblocks to print pictures and cards ever since paper had become widely available, and on fabric before that.

However, Gutenberg's printing process combined several important innovations. The first was making individual letter types, which could be reused and recombined to form any sequence of text. The second was making them out of cast metal. These could be mass produced from a single matrix and were far more durable than wooden or ceramic types. When they wore out they could be recast again and again from the original matrix. The third innovation was using a new sort of ink that sat on top of the paper like varnish. Finally, no one had used a screw mechanism to press down the paper before. When put together these apparently minor improvements created a machine that proved to be revolutionary.

Gutenberg, and later Fust and Schoeffer, tried desperately to keep the invention to themselves, swearing workers to secrecy and describing their discovery obscurely as the Work of Books.[6] At first they didn't even let on that they had a new method for producing books. One story reports Johann Fust in Paris trying to pass off printed Bibles as manuscripts. Some of the people he was trying to sell to noticed how uniform the letters were and accused him of conspiring with the Devil. He was forced to admit that the books were machine-made or face prosecution for heresy.[7]

In any case, their efforts were futile. Political turmoil in Mainz in 1462 led many experienced printers to leave town, and from this point printing spread rapidly. As a new industry, and unlike most occupations, printing fell outside the control of the powerful trade guilds, the Church, or the State so there were none of the usual constraints on setting up a print workshop. Presses sprung up along the trading routes of the Rhine, then moved east across the Alps into Italy and west to Strasbourg and Paris. Printing reached the Netherlands and England in the 1470s, and by 1480 there were printing presses in at least 110 European towns.[8] Printing had spread to every major urban centre in Europe by the end of the century.[9]

Gutenberg's modest aim was to make copying books more efficient, and in this he succeeded. Printing made books cheaper, more plentiful, and less prone to scribal error. But that apparently simple outcome triggered consequences that resonated far beyond the book trade.

Printing didn't just change the way books were made, but the way they were read, who read them and what they contained. It changed how people taught and how people learned, how they wrote and how they remembered, how they saw the past and how they saw the future.

These changes didn't happen overnight, nor were they straightforward. The two centuries after the invention of the printing press were characterised by major conflicts, confusion and trauma – religious, academic, political and psychological. The Latin Church, dominant institution of the previous millennium, was discredited and broken into pieces by the

Reformation and Counter-Reformation, at the cost of thousands of lives. Scholarship was turned on its head as scholars, faced with unprecedented access to new texts, discovered shortcomings and inconsistencies in wisdom once taken as unassailable. Most importantly, right across Europe the principal means by which knowledge and information were transmitted shifted from the spoken to the written word: from the ear to the eye.

Europe, Fifteenth Century

To appreciate the impact of the printing press we need to reflect on the society into which it was born. Despite the fact that the alphabet had existed for 2,000 years, and other scripts for millennia before that, most people were still illiterate in the fifteenth century. Outside the Church and the universities, Europe was still a traditional society in which the vast majority of knowledge and information was passed on directly from person to person in spoken words and actions. Day-to-day life was noisy, smelly, tactile and active. Towns were congested, bustling places populated by people and animals constantly on the move. Artisans laboured in hot, noisy workshops that spilled onto narrow streets that reeked of animal and human excrement. In the markets, merchants loudly hawked their wares while buyers squeezed, weighed and sniffed them, asked questions, and haggled over prices. Deals were concluded with a handshake as they had been since Babylonian times.[10] Troupes and troubadours, mummers and minstrels performed poems, plays, ballads and songs in crowded town squares and taverns.

Bells and horns sounded the hours to signal the beginning and end of the working day, calls to prayer and evening curfew. When mechanical clocks were first invented they had no faces or hands; they struck the hours on a bell as a human bell-ringer had previously. Proclamations, news and announcements were delivered in person by town criers. In many towns rear watchmen patrolled the streets throughout the night, lantern in hand, singing out a rhyme that told the hour and the weather, and loudly bidding the townsfolk good rest.

Law and order were person-to-person matters despite centuries of written legal records. Open-air courts heard verbal testimony given under sworn oaths, generally held to be more credible than written documents. In English law, in a process called compurgation, a defendant could assemble a prescribed number of 'oath helpers' to swear before the court that he was telling the truth, in which case the plaintiff's case would necessarily fail. Certain cases were still subject to trial by ordeal, where a defendant's guilt or innocence depended on their reaction to a physical ordeal such as being thrown into a fire or deep water.

Religion was also conducted out loud and in person. Wealthy people had prayer books, but private Bible ownership was practically unknown, and translating the Bible from Latin into vernacular languages was illegal.[11] Church attendance was an immersive, crowded, multisensory experience of music, chanting, incense, candles, light streaming through stained glass, sculpted columns and frescoed walls. It was also mobile, as there were no chairs in churches, and people would mill around during Sunday services waiting for the moment when the priest raised the host – the consecrated bread that had become the body of Christ – high above his head, conferring a blessing on all who saw it.

Learning happened almost entirely by doing. Children mostly learned from helping their parents at home, in the fields or in the workshops and marketplaces. Urban boys and girls attended elementary school, where they learned the basic rules of grammar and the prayers and lessons known as the Catechism. After elementary school, boys joined their fathers in the field, were apprenticed to a master to learn a trade, became page to a knight, or joined a seminary. Some went on to grammar schools with a view to university, but these were a tiny minority. Girls learned cooking, weaving, embroidery and other domestic skills in preparation for marriage. Very little was written down.

Today, oral cultures have all but disappeared, but in the last century anthropologists were still able to study preliterate societies, untouched by the West.[12] Researchers have discovered that in these cultures life was lived through all the senses, but primarily through hearing. Rituals, routine and religion helped people remember and enforce their society's rules for communal living. Interpersonal relationships were critical, as life was inherently intimate, shared, and personal. Everyone knew their place in the greater organism of the community, and the strict system of rules meant there was little need for personal initiative or individual responsibility. Life was lived in a present that included both the past and the future, not as a linear sequence but as an organic experience where bodily life and death were simply different states of the eternal reality of the soul.[13] Nature's workings tended to be explained by reference to higher powers – a god or gods (good or bad), or ancestors – or accepted as beyond the ken of mere mortals.

Fifteenth-century Europe shared many of these chara-cteristics, despite the long presence of writing and a partially literate population. A strict set of rules – some codified, some set by custom – dictated how people lived and worked, who could go where, who could do what, and what they were supposed to believe.

Our language is peppered with the remnants of Europe's former oral culture. Our word *clock* comes from the French word *cloche*, meaning bell. Contrast this with the silent, individual *watch*, invented a few years after the printing press. The word *catechism*, which now describes religious learning generally, comes from the Greek for learning orally. We call court proceedings *hearings*, and we approve of something with the phrase *Hear! Hear!* In business, financial statements are *audited*, from the Latin verb to hear, because in medieval England financial records were checked by having them read aloud.[14] We call those who experience a performance the *audience*, from the same Latin root, and Hamlet declares 'we'll hear a play'[15] in the Shakespeare play written in 1600, even though by this time most people were already saying 'see' a play and using the term 'show' interchangeably with 'play'.[16]

In the absence of books, medieval communities employed a variety of tools for retaining and transmitting important information. Rhythm and rhyme were used everywhere, not only in storytelling and entertainment but also for remembering everyday practicalities, trade practices, even laws and regulations. A twelfth-century Irish document recorded the rights, taxes and stipends of the Irish kings and lower nobles entirely in rhyme.[17] Spells and incantations were also commonplace, not to conjure magic but as aids to remember recipes or other everyday necessities. The word *spell* in Old English meant 'story' or 'speech' and was only later used to describe occult practices.[18]

Even among scholars and clergy – by far the most literate medieval communities – speaking was much more important than reading and writing. Teaching and preaching happened aloud, from memory. This had been the case since antiquity, when the Art of Rhetoric – the word comes from the Greek word *rhetor*, meaning speaker – was developed. All educated people learned the Art of Rhetoric in the Middle Ages, along with the rules of language known as grammar, and the principles of logic.[19] Wisdom was equated not with accumulated knowledge but with eloquent speech. Some of the greatest teachers and preachers of all time, including Socrates, Pythagoras and Jesus Christ, never wrote at all,[20] and modern scholars believe much of what was 'written' by the great thinkers of antiquity was never actually written by them, but rather was captured from what they said in notes made by their followers and students.

People weren't taught literature because it didn't really exist as an art in its own right. The late philosopher Walter Ong, who wrote extensively on the shift from oral to literate culture, described writing prior to the printing press as a device for recording the spoken word rather than an art of its own.[21] Manuscripts recorded, stored and played back words that were said or sung; their purpose was to transfer the speaker's words to the reader's memory.

This purpose was reflected in the way books were written and read, and in their design.

Whether manuscript (handwritten) books were scholarly, theological or entertaining, they were constructed as if written for an audience, as they had been since antiquity. Most writing reflected verbalised forms: theological works read (sounded) like a sermon, a confession or a prayer; scholarly works read like a lecture or letter. Books were also rarely actually written by the author but more often dictated to a scribe or an amanuensis.* One can readily imagine the learned man pacing the room, pronouncing his deliberations to an imagined audience, while a scribe struggles furiously to capture his words.

Reading was usually done aloud, in company, the audience was supposed to imagine a speaker or speakers there with them in the room. Reading aloud for entertainment remained

---

* The most revered biblical author, St Paul, didn't actually write his famous letters but dictated them to his amanuensis Tertius, as is clear from Romans 16.22.

popular well into the eighteenth century, in part because of lingering illiteracy but also because candles were expensive. Even lone reading was often done out loud. Reading was not intended to be a rapid relay of information; rather it was meant to be something of a 'meditative meandering',[22] analogous to a child's relationship with a much-loved storybook, where the story is read aloud again and again until both reader and listener can recite it word for word.

Manuscript design gives a further clue to literacy's hybrid nature before printing. One of the most distinctive features of medieval manuscripts was their use of illumination – the tiny pictures and graphic designs and lettering that adorned their pages. Illumination was part of a reading process whereby each page was intended to be pored over, considered, studied and, ultimately, remembered. Text and images combined to help the reader in this task.[23]

Recent research suggests manuscript illumination may have played a very specific part in the reading process. Scholars now believe that illuminations served as memory aids for readers, both as markers to help readers locate a particular part of the text, and as visual pointers that readers could use to create memory 'hooks' on which to hang particular pieces of information.[24] According to this theory, books were a particular form of Memory Palace, an ancient device for memorising long or complicated strings of information that formed part of a wider mnemonic system known as the Art of Memory, a subset of the Art of Rhetoric. The Art of Memory was invented by the Greek poet Simonedes around 500 BCE and endorsed by the great Roman orator Cicero.[25] It was rekindled in the Middle Ages, in particular by St Thomas Aquinas (1225–74), as a way of remembering the complex hierarchy of sins, virtues, vices, rewards and punishments that were so important to medieval theology.

The Art of Memory is a visualising technique based on things and places. The first step is to construct a physical locality in your imagination, such as a large building – hence a Memory Palace – or a streetscape or town. Next, create a memorable visual image to represent each thing to be remembered. Images are easier to remember, apparently, if they are 'base, dishonourable, unusual, great, unbelievable or ridiculous' rather than 'petty, ordinary and banal'.[26] With imaginary images in hand, travel around the imaginary space and place each image, one by one, somewhere within the space. In the *first place* goes the first image, in the *second place* the next, and so on. To recall the information later, make the same imaginary journey around the space, *re-collecting* the images one by one. Modern language carries the echoes of the practice to this day.

The idea of using a series of strangely memorable images as a memory device casts medieval visual art in a new light. The fantastical creatures, bizarre characters and apparently random birds, animals and other images that one encounters in medieval art of all forms are suddenly infused with a new sense of purpose. Perhaps they are not random at all: they help illiterate people remember. Art historian Michael Baxandall makes the further suggestion that the reason people and places were often painted rather generically in the late Middle Ages may have been so that they could be a memory pallette on which people could impose their own mnemonic detail.[27] The contrast between the bland faces of many medieval figures and the bewildering strangeness of much of medieval art may, in fact, reflect a very deliberate effort to provide illiterate people with tools for remembering the complex rules by which they lived and worshipped.

If this all sounds extremely convoluted, I couldn't agree more. But the Art of Memory was taken extremely seriously in the Middle Ages and well into the Renaissance. Versions of the now all-but-forgotten technique are still used today by so-called mental athletes training for memory competitions, where they perform such astounding – but not very useful – feats as remembering the order of appearance of a large number of playing cards, or random sequences of words.[28]

Western Europe, 1450–1500

From the outset, printing was a business enterprise and, like publishers today, printers competed for content and customers. The first printed books were aimed at the clerics and scholars who had always made up the market for manuscripts. Bibles and other religious works were closely followed by academic texts, especially the classical Greek, Roman and Arabic works beloved of the Humanist movement that was burgeoning in Italy and beginning to take hold elsewhere.

As the printing trade spread and became more competitive, printers started looking for ways to distinguish their own products. The earliest printed books were – quite intentionally – almost indistinguishable from manuscripts, so their typefaces closely resembled handwriting. Gutenberg's Bible, for example, used heavy Gothic blackletter script, and left spaces for hand illumination to be added after printing. As printed books became widely accepted, printers made efforts to make text more legible.

Around 1464 a Strasbourg printer cut a set of letter types based on the Carolingian Miniscule style invented by Alcuin centuries before. Then in 1470 the Venetian printer Nicholas Jenson created a new typeface designed specifically for print, combining Carolingian Miniscule with Roman-style capitals. Jenson's Roman typeface, as it was called, eventually became standard across Europe, down to the present day; you are probably reading a version of it now.[29] The exception was in

Germany, where many printers retained a Gothic script called Fraktur right up to the twentieth century, believing it to be more Germanic than the Latin-influenced Roman script. Fraktur is the typeface we now associate with Nazi propaganda, although, for reasons unknown, midway through the Second World War Hitler's private secretary Martin Bormann issued an order, headed in Fraktur type, that all typeface should hence-forth switch from Gothic Fraktur to Roman. The Gothic style, he said, was associated with Jewish interests.[30]

Printers also started adding illustrations from carved wood-blocks to printed texts. At first they were printed by hand onto spaces left in the text, like a rubber stamp, but printers soon fig-ured out a way of incorporating woodblock and typeface into a single printing process. As illustrated books grew more popular, ambitious printers commissioned serious artists to design them. Hans Holbein the Younger, Albrecht Dürer and Pieter Breugel all designed for printed books.

In 1472 a Veronese printer produced the first book with tech-nical – as opposed to religious or decorative – illustrations.[31] *De Re Militari* pictured weapons, war chariots, siege engines, can-nons, flags, water floats, bridges and pontoons, and much more besides.[32] Later that year, the first map appeared in a printed book, and five years later Ptolemy's World Atlas was published.

Printers commissioned scholars to recreate accurate rendi-tions of charts, maps, drawings, tables and diagrams, many of which had been impossible to reproduce accurately by hand. Printed illustrations provided incalculable value to students of anatomy, architecture, astronomy, alchemy, botany, mathemat-ics and all manner of other fields. An early botanical guide had 131 illustrations named in Latin, Greek, Persian and Egyptian,[33] while an edition of Euclid's *Elements* contained more than 400 geometric figures. In 1483 the first set of mathematical and astronomical tables, the *Alphonsine Tables*, was printed.[34]

Another major innovation came about in 1474 when a history of the world was the first printed book to contain page numbers.[35] This was by no means a trivial improvement. It was almost impossible to identify or reference a particular piece of information in a manuscript without providing a raft of cumbersome detail that relied heavily on memory. Once printed books were paginated, scholars could refer colleagues directly and succinctly to a particular page of a particular edition, and similarly refer back themselves.[36] Pagination soon led to contents pages and indexes, and cataloguing using novel systems such as chronology and alphabetical ordering. While these methods may seem obvious now, medieval scholars and librarians each had their own idiosyncratic methods of cataloguing written material, once again relying heavily on memory. The eighth-century Carolingian scholar Alcuin, for example, catalogued his own library in rhyme.[37]

By the early 1500s the great medieval art of illumination was all but dead,[38] replaced by the detailed and precise diagrams and illustrations printing could support. A printed book no longer needed to be a tool for memory. Once books were in ready supply and easy to navigate, there was no need to memorise them. Books were increasingly created for reference, rather than memory. They were no longer companions or works of art but tools. From that time the great works of painted art were to be found not in books but almost exclusively on walls, panels and a new medium, conveniently portable canvas.

Along with these presentational and organisational improvements, printers set about correcting the many errors and inconsistencies that had crept into manuscripts as they had been copied and recopied over time. Scholars and clerics had long bemoaned the ignorance and sloth of copyists who 'spoil everything and turn it into nonsenses'.[39] This had only worsened as booksellers in some of the growing university towns attempted to mass produce manuscripts using large teams of scribes.[40] The printer, freed from the effort of handwriting, could concentrate on the content and arrangement of the text on the page. This was no mean feat. One anonymous scholar correcting a text described how he took great pains in the correction of it:

> I have sought out diligently all the copies which I have been able to discover for this purpose in any of the libraries in the school of Heidelberg, in Speyer and in Worms, and finally also in Strasberg. And since in the course of this I have learned by experience that that particular book ... is rare to come by even in the great and well stocked libraries, and even rarer can it be had for copying from any of those same libraries; and also, what is worse, that when it can be found in there it is more rarely corrected or emended; on that account I have been moved to work most carefully to this end.[41]

The Consolidation of Knowledge

Medieval scholars knew all too well that knowledge could be lost forever. It had happened several times before with the destruction of the great library at Alexandria, after the fall of Rome, again after the brief Carolingian Renaissance, and with the destruction of the House of Wisdom in 1258. It is therefore not surprising that a large proportion of scholarly effort in the fourteenth and early fifteenth centuries was devoted to preserving the contents of the ancient texts rediscovered over the previous few centuries.

The printing press fundamentally redirected this scholarly focus. Once dozens of copies of a book were printed and distributed, its contents were secure. The past became indelible, freeing scholars to devote their time to improving the body of knowledge rather than merely sustaining it. One of their first tasks was sifting through the hundreds of texts printers were gathering up and printing, and figuring out what was truly valuable wisdom or technical insight and what was bogus, erroneous, or simply nonsense. Scholars could suddenly access dozens or even hundreds of texts at once, and unsurprisingly a great deal of confusion ensued between different versions and translations of a particular text (not least the Bible), conflicting theories, outlandish claims, forgeries and fakes. They realised that not every work handed down from the past was to be revered, and nor was every ancient work necessarily ancient.

The process of cataloguing, collating, comparing, critiquing and debating led to a newly critical approach to knowledge and wisdom, and a new spirit of intellectual challenge and enquiry that was the precursor to the modern mindset.

Seeing Secrets

After the initial tranche of books aimed at scholars and clerics, printers set their sights on a new customer base: the urban middle classes. Printers targeted the new bourgeoisie of merchants, tradesmen, artists and artisans with well-known books translated into vernacular languages and made newly affordable. These included chivalric romances, poems, almanacs and calendars, books of hours and psalms.

Before long, printers began seeking out new material for the middle-class audience. A text on arithmetic for merchants was printed in the Venetian vernacular in 1478, and this was followed in the next decade by printed tables for calculating profit shares, interest, currency exchanges and so on. Before the end of the century the Italian mathematician (and teacher of Leonardo da Vinci) Luca Pacioli published a mathematical text that included a description of double-entry bookkeeping. This may not be the most obviously sexy topic but the book has been described as the 'most influential work in the history of capitalism'[42] and reportedly transformed Renaissance commerce.[43]

Thereafter came a raft of what we might call general interest and 'how to' books: travelogues, cookbooks, guides to etiquette, household tips, remedies and recipes. As time went on, the territory these books covered extended into more technical subjects such as alchemy, metallurgy, surgery and other aspects of what we would now call popular science.

These subjects were not considered sciences at the time. They were of no scholarly interest, just everyday topics belonging to the underworld of learning inhabited by ordinary people.[44] They included the vast miscellany of traditional and practical knowledge that ranged from the sage to the silly, taking in all the skills, trades and crafts that underpinned medieval life.

What these subjects had in common was that until they were captured in print they were, for the most part, secrets.

We take for granted the idea that knowledge should be celebrated and shared, but the medieval attitude was completely different. Most of what was known was locked tightly into closed circles, passed along from person to person within a particular circle in the expectation that it would never leave it. Academics and religious orders used Latin to help keep their secrets from the vulgar masses, but they also resorted to obscure language, codes and ciphers to keep their knowledge secure.

The paradigm of secrecy prevailed elsewhere too. Below the clerics and scholars, artisans, tradesmen and merchants formed their own exclusive circles. The guilds and liveries were just as protective of their own secret knowledge as the elite classes.[45] Other circles were informal but equally secretive about their arts, such as midwives. Then there were the covert circles that existed outside mainstream society and dabbled in the occult and magic. Secrets were kept at every level of society.

Inherent in the notion of secrecy was the idea that certain knowledge should be kept from those unworthy souls who might misunderstand or misuse it. Specialist knowledge was universally regarded as a sacred gift from God, revealed only to those in His confidence, and those privileged to have access must therefore take great pains to respect that confidence.

Curiosity was actively discouraged – an injunction that went back to Adam and Eve in the Garden of Eden. Secrets were not meant to be probed, and even apparently minor transgressions could be punished. A Belgian court fined a man 15 livres for hiding behind a staircase to eavesdrop on his wife during child-birth. The court found such curiosity 'doth not befit a man'.[46] Divulgers of secrets were also treated harshly. Some craft guilds were so protective of their trade secrets they would track down and kill any member who left their home city, rather than risk indiscretion. Church law decreed that a priest who revealed what he had been told in confession would be permanently banished to penance in a monastery.[47]

Counterintuitive as it may seem, secrets were actually a fairly efficient way of preserving knowledge in a mostly preliterate culture. Distributing specialised knowledge into pockets kept the volume of things any one individual had to remember at a manageable level. Proclaiming its secrecy within a tight, exclusive circle kept the knowledge precious and reduced the chance of it being forgotten or carelessly lost.

Nonetheless, along with the extreme compartmentalisation of knowledge came stagnation. Artisans had practical knowledge but knew nothing of the theory behind their craft. Theory was the exclusive privilege of academics, who likewise had no access to the practical wisdom that might allow them to apply those theories in useful ways. As these two classes were unlikely ever to encounter one another, for centuries they each carried on in happy ignorance of what they might learn from the other.

The printing press entered this complex network of mysteries and secrets and blew it apart. A printed book is the very opposite of a secret. Every copy is identical to 1,000 others and can spread as far and as fast as a printer's horse or merchant's ship can carry it. Printers and booksellers are secrets' worst enemies: they want to print and sell as many copies as they can. And what sells better than a secret?

Thus the sixteenth century saw a new genre, 'Books of Secrets', that promised to reveal ancient mysteries and wisdom. Written in everyday vernacular languages, they offered practical prescriptions, explanations and recipes, often with diagrams and illustrations. Some presented simple home truths. Others, such as *The Secrets of Lady Isabella Cortese* – a mysterious (and possibly fabricated) Italian noblewoman – described complex alchemical processes requiring sophisticated equipment and ingredients that wouldn't be out of place in a chemistry laboratory.[48] They highlighted their secrecy with flowery – and of course entirely illogical once printed and reproduced – exhortations to the reader to keep the secrets therein, and 'guard it as he would his own soul and not to give it to any stranger'.[49]

By revealing the secrets of the practical and mechanical arts, the printed books of secrets dissolved the invisible walls that had held medieval knowledge captive for centuries and released it into the world. Expertise and information that had only ever existed in the heads and hands of select initiates were made visible for the first time, put down on paper in text and diagram, where they were laid bare to scrutiny and analysis.

Inevitably, theory and practice began to intermingle, both in the minds of readers and writers and in printers' workshops, where authors from all walks of life might meet. New connections were made, generating radical new ideas. Many of nature's 'secrets', it turned out, were not secrets at all; the answers had just been hidden from view. People began to believe further investigation would reveal more answers and started actively hunting for them. A new type of person appeared: the self-taught scientist, an enthusiastic amateur outside the established academy who pursued knowledge for his or her own curiosity or, perhaps, with dreams of the glory of discovery. The Renaissance Man was born.

The ancient injunction against probing the divine Secrets of Nature remained, but was finally lifted in the late sixteenth century when the influential philosopher Francis Bacon distinguished between natural and religious secrets. The scriptures warned against 'the ambitious and proud desire of *moral* knowledge to judge of good and evil'.[50] On the other hand, Bacon pointed out, the pursuit of natural philosophy, or science as we now know it, was not a threat to religion. It was, he said, 'at once the surest medicine against superstition and the most approved nourishment for faith'.[51]

Nature, for centuries a divine and forbidden mystery to be wondered at, became instead a fascinating puzzle to be solved by the emerging new art of science. This was the beginning of the modern mindset, a completely new way of seeing the world, and one that could never have come about without Gutenberg's printing press.

Wittenberg, 1517

On 31 October 1517, a young priest and scholar named Martin Luther posted a letter to his bishop on the door of the All Saints Church in Wittenberg. The letter contained ninety-five points questioning various Catholic Church activities, in particular the sale of Indulgences. These were the 'get out of jail free' certificates people could buy as penance for their sins or to reduce their time in purgatory. Gutenberg himself had printed the first mass-produced Indulgences in 1454 while he was working on his Bible project, and by the end of the fifteenth century they had become an untrammelled money-making racket for the Church. At the time of Luther's protest, Indulgences were being sold at an unprecedented rate to raise money for the rebuilding of St Peter's Basilica in Rome.

Luther's action in posting his letter was a conventional method of initiating an academic debate at the time, but fellow dissidents recognised a wider opportunity. They saw that the printing press could take his scholarly salvo beyond the walls of the bishop's palace and into the popular consciousness. Luther's supporters turned the '95 Theses' into a pamphlet and printed and distributed it in towns across Germany. Luther had written his theses in academic Latin, but reformers translated them into vernacular German and other languages, and illustrated them with satirical cartoons and caricatures. Their propaganda campaign proved highly successful. Within a couple of months, the Theses had spread all around Europe, triggering the movement that became the Protestant Reformation – an event that played out over the next century and a half and fundamentally changed religious life in Europe.

Once the reforming message was broadly dispersed, it was impossible to dislodge. Sheer weight of numbers reached by the printed pamphlets and the indelible record of reams of printed material meant the Church couldn't gather up Luther and his heresy and burn them at the stake as it had previous attempts at reform. The Protestant Revolution, writ large in print, could not be quashed.

The initial spread of Luther's Theses could not have happened without the printing press. But printing and the Reformation were entwined in other ways that reflected the fundamentally different views Protestants and Catholics had on the role of seeing in religion. For centuries in the Catholic Church, rich imagery was both a way of glorifying God and also a tool for remembering the stories and lessons of the Church. The Catholic cathedral and all it contained were the liturgy of the leity, the Bible of the illiterate. For Protestants, the written words of the scriptures were the only legitimate portrayals of God, and any other imagery risked idolatry. Reformers destroyed countless religious paintings, sculptures and other works of art in their zeal to obliterate what they saw as Catholic idolatry, as the eyewitness account of an outbreak of iconoclasm in the Netherlands by one Richard Clough describes:

> We have had here this night past a marvellous stir, all the churches, chapels and houses of religion utterly defaced, and no kind of thing left whole within them, but broken and utterly destroyed, being done after such order and with so few folk that it is to be marvelled at … they began with the image of Our Lady, which had been carried about the town on Sunday last, and utterly defaced her and her chapel, and, after, the whole church, which was the costliest church in Europe … and coming into Our Lady church, it looked like a hell, where were above ten thousand torches burning, and such a noise as if heaven and earth had got together, with falling of images and beating down of costly works, such sort that the spoil was so great that a man could not well pass through there.[52]

In England, iconoclasm was officially sanctioned by King Henry VIII's successor, Edward VI. The 1537 *Chronicle of the Greyfriars* records that:

> In September began the King's visitation at [Saint] Paul's and all the images pulled down: and the ninth day of the same month the said visitation was at St. Bride's, and after that in

divers other parish churches; and so all images pulled down through all England at that time, and all churches new white-lined with the commandments written on the walls.[53]

Protestants were determined to replace pictorial images with images of words: the written Bible. While lay Catholics were forbidden from owning Bibles or reading them in their own languages, personal Bible reading and ownership were central aims of Protestantism. Protestants rejected the notion of the church as Bible of the illiterate and instead set about teaching people to read. Henry VIII authorised the preparation and printing of an English language Bible in 1538 (despite having burned the author of an earlier English version at the stake a decade earlier) and ordered the clergy to place a copy in every church in the land in a place where parishioners could come and read it.[54] A year later the various local forms of Mass were replaced by law with a universal service printed in the Book of Common Prayer. None of these Protestant reforms would have been conceivable before printing.

Protestants replaced the multisensory, three-dimensional worship of the Latin Church with two-dimensional, printed scriptures and prayer books. Their intention was acutely pious: the Reformers passionately believed their way of worshipping God was correct. However, I wonder whether, in rejecting the immersive religious experience in favour of a single channel of communication with God (the written word), the Protestants subtly removed religion from the daily lived experience of congregants, and in so doing hastened the separation of religious and secular life.

Manuscript books augmented oral culture. Printed books replaced it. While printing's impact was initially quantitative, increasing the number of books made, the nature of the changes it brought about were ultimately qualitative, changing the very nature of thought, communication and society. A century and a half after the first printed pages came off the press, Europe had changed, fundamentally and forever.

The printing press wasn't, strictly speaking, a visual invention. It simply mechanised an activity that had been going on for thousands of years. But the invention of the printing press was a seminal moment in the history of seeing. By making books and other printed materials much more accessible, printing introduced reading and writing into every facet of daily life. Learning, work, society, citizenship and religion were all deeply affected as entrepreneurial printers found more and more opportunities to capture knowledge and communication in written (by which I mean printed) form. It erased the substantially oral culture that had existed for all of human history, and still characterised Western Europe in the fifteenth century, and transformed it into the written culture that we take for granted today. And while oral culture relies on all the senses, written culture is, overwhelmingly, visual.

Within two centuries, the printed page became the predominant medium for the exchange and dissemination of knowledge and information in Western society. Knowledge became cheap, portable, and almost endlessly reproducible. Meanwhile, a substantial proportion of daily human experience was converted from multisensory and personal encounters into the detached, private scrutiny of a silent printed page.

The printing press rendered all forms of knowledge visible. It broke open the cells that had kept much of what was known secret for centuries and turned that oral and practical wisdom into written words and pictures. In a curious twist, printing made the practical arts literate and the literate sciences practical.

It changed our minds in more subtle ways, too. A few generations into living in a world of print, Western consciousness took on some of the characteristics of printed text. Logical, detached, rational thinking replaced faith in God's mysteries; enquiry replaced awe. The experience of time became linear rather than cyclical; standardisation and repetition replaced idiosyncrasy and craft. Clarity replaced mystery.

Above all, printing channelled our senses into our eyes, making them the primary source of learning, entertainment, knowledge and information for the foreseeable future. We haven't looked back since.

10

THE EYE, EXTENDED:
THE TELESCOPE

Venice, 1610

THE STARRY MESSENGER
Revealing great, unusual and remarkable spectacles
Opening these to the consideration of every man
And especially of philosophers and astronomers;
As observed by Galileo Galilei
Gentleman of Florence
Professor of Mathematics at the University of Padua,
With the aid of a
SPYGLASS
lately invented by him,
In the surface of the Moon, in innumerable Fixed Stars, in
Nebulae, and above all in
FOUR PLANETS
Swiftly revolving around Jupiter

Title page in Galileo Galilei,
The Starry Messenger, 1610

Rome, 1616

The whole Congregation of the Holy Office, ordered and enjoined the said Galileo, who was himself still present, to abandon completely the above-mentioned opinion that the sun stands still at the center of the world and the earth moves, and henceforth not to hold, teach, or defend it in any way whatever, either orally or in writing; otherwise the Holy Office would start proceedings against him. The same Galileo acquiesced in this injunction and promised to obey.

Special Injunction of the Roman Inquisition, February 1616[1]

This Holy Congregation has also learned about the spreading and acceptance by many of the false Pythagorean doctrine, altogether contrary to the Holy Scripture, that the earth moves and the sun is motionless, which is also taught by Nicholaus Copernicus's On the Revolutions of the Heavenly Spheres … Therefore, in order that this opinion may not creep any further to the prejudice of Catholic truth, the Congregation has decided that the books … be suspended until corrected; … and all other books teaching the same thing are likewise prohibited.[2]

Holy Congregation of the Index of
Forbidden Books, March 1616

Convent of Minerva, Rome, 1633

I, Galileo … swear that I have always believed, do believe, and by God's help will in the future believe, all that is held, preached, and taught by the Holy Catholic and Apostolic Church … I have been pronounced by the Holy Office to be vehemently suspected of heresy, that is to say, of having held and believed that the Sun is the centre of the world and immovable, and that the earth is not the centre and moves:

... with sincere heart and unfeigned faith I abjure, curse, and detest the aforesaid errors and heresies, and generally every other error, heresy, and sect whatsoever contrary to the said Holy Church, and I swear that in the future I will never again say or assert, verbally or in writing, anything that might furnish occasion for a similar suspicion.

I, Galileo Galilei, have abjured as above with my own hand.

Recantation of Galileo, Roman Inquisition, 1633[3]

Rome, 1822

The printing and publication of works treating of the motion of the earth and the stability of the sun, in accordance with the opinion of modern astronomers, is permitted.

Statement of the College of Cardinals, Vatican, 1822

Rome, 1835

Galileo's *The Starry Messenger* and Copernicus' *De Revolutionibus* are removed from the Index of Forbidden Books.

Rome, 1992

A tragic mutual incomprehension has been interpreted as the reflection of a fundamental opposition between science and faith. The clarifications furnished by recent historical studies enable us to state that this sad misunderstanding now belongs to the past.

From the Galileo affair we can learn a lesson which remains valid in relation to similar situations which occur today and which may occur in the future ...

Often, beyond two partial and contrasting perceptions, there exists a wider perception which includes them and goes beyond both of them.

Another lesson which we can draw is that the different branches of knowledge call for different methods ... The error of the theologians of the time, when they maintained the centrality of the Earth, was to think that our understanding of the physical world's structure was, in some way, imposed by the literal sense of the Sacred Scripture. Let us recall the celebrated saying attributed to Baronius: 'the intention of the Holy Spirit is to tell us how to go to Heaven, not how the heavens go' ... There exist two realms of knowledge, one which has its source in Revelation and one which reason can discover by its own power ...

Pope Jean Paul II, Address to the
Pontifical Academy of Sciences, 1992[4]

New York, 1992

After 350 Years, Vatican Says Galileo Was Right: It Moves

Front page, *New York Times*, 31 October 1992

Twinkle, twinkle little star,
How I wonder what you are?

Is it possible to look into a clear night sky and not be struck with wonder and awe? Humankind has asked the simple question posed in the famous nursery rhyme again and again throughout recorded history, and probably for thousands of years before that.

From the earliest times, people realised the night sky is not a static canopy. The stars sweep magisterially across the sky, rising in the east and setting in the west like the Sun, rotating slowly around a single point. Some cultures realised the changes in the night sky had rhythms and patterns that rendered its movements a form of language that could turn the sky into calendar, map and compass – thousands of years before calendars or map-making or compasses existed. Some believed the night sky's silent language foretold fates and destinies, and structured their lives and decisions according to its movements. Many tried to decipher its secrets. Philosophers theorised and mathematicians calculated, but for millennia the stars remained enigmatic, visible yet invisible, unreachable and unknowable.

Then, in the first decade of the seventeenth century, a humble spectacle-maker presented an invention that would eventually provide some answers to the age-old question. It would also ignite a revolution in science that rocked the very foundations of European society, and raised fundamental questions about truth, belief and Christianity that are still being debated today.

In doing so it raised new and even more awe-inspiring questions, many of which continue to tax some of greatest minds on the planet. The 1609 invention of the telescope shoved mankind off his pedestal at the centre of the universe, even while giving him greater dominion than ever over his own tiny planet and the Secrets of Nature.

King Sun and Queen Moon

Not many of us today would change our plans in response to a newspaper horoscope, but we are ruled by the motions of the sky in more ways than you may think. The Sun's daily and annual cycles affect every living thing on Earth, from the lowliest bacteria to the highest mammal, including humans. Internal circadian (daily) clocks affect all aspects of life ranging from overt behaviours like sleep, feeding and reproduction to invisible biological functions such as the release of hormones. Multiple studies have linked night workers to higher levels of mental illness and other health problems, demonstrating how reliant we are on our primal rhythms.[5]

The seasons – governed by the Sun – affect life profoundly, be it trees losing their leaves in autumn, mammals growing a winter coat or birds migrating across the world in the spring. We humans feel the change of seasons no less than any other species and, like every species, structure our lives around the annual solar calendar.

Many animals also respond to the cycles of the Moon.[6] The female species of a certain crab spends most of her adult life living in mountains but uses the Moon's cycle to determine when to come down to the shore to release her young into the sea. Hundreds of species of coral spawn simultaneously in response to a lunar trigger, while in West Africa millions of *Povilla* mayflies emerge from their pupal cases exactly two days after a full Moon, at which point they have only an hour or two in which to perform their mating flight, mate and lay their eggs before they die. Some bird species hatch during a new Moon so that a couple of weeks later, when the newborns are most demanding, the full Moon makes it easier to find food for them.[7] Contrary to popular belief, however, and despite the ubiquity of this ancient and iconic image, there is no scientific evidence that wolves habitually howl at a full Moon.[8]

We may not be aware of it, but humans are also deeply affected by lunar cycles. Hospital admissions, births, accidents and suicides have all shown correlations with Moon phases.[9]

Ten thousand years ago, Stone Age nomads living in Scotland built a series of twelve pits, arrayed in an arc. Seen from the sky, the pits mimic the stages of the lunar cycle. The shape of the arc, meanwhile, aligns with a notch in a hill on the horizon through which the Sun rises on midwinter's day, marking the annual winter solstice. Researchers believe the structure is a calendar: the oldest marker of time ever discovered anywhere. They believe it was used to estimate the timing of game migration and salmon runs (it pre-dates settled farming in Britain by several millennia), and may also have played an important spiritual role.[10]

Remarkably, these Scottish hunter-gatherers had realised that the lunar and solar cycles don't align. A lunar month takes twenty-nine days while a solar year takes 365 days, or twelve and a third lunar months. A lunar-based calendar thus slips from the annual solar cycle by about eleven days a year, taking it out of line with the seasons.* The Scots seem to have recognised that knowing the date of an annual solar event such as midwinter allowed the lunar calendar to be recalibrated with the solar seasons. Evidence at the Scottish site indicates the calendar was in use for several millennia – quite a feat for people who used only stone tools.

---

* Jewish and Islamic calendars are based on lunar cycles, which is why festivals such as Ramadan and Rosh Hashana occur at different times each year. Christian Easter is also set based on lunar cycles, which is why its date also moves around.

The Starry Night Sky

Along with the Sun and Moon, many ancient cultures looked to the stars to determine the passage of time. At least 5,000 years ago Egyptians knew that when the brightest star Sirius became visible, just before dawn, the Nile floods were coming. Similarly, Australian aborigines knew that when the Pleiades cluster appeared in the evening sky, spring had begun.[11]

The Sumerians of ancient Mesopotamia started observing, measuring and cataloguing the night sky at least 4,000 years ago. They saw that the canopy of stars rotated overhead every twenty-four hours. They observed that while most of the stars remained in fixed positions relative to one another, a few stars travelled around within the fixed pattern. They believed these 'vagabond' stars were gods and named them after their deities. The Greeks later called this group of stars the wanderers – *planetes* – and named them for their gods Mercury, Venus, Mars, Jupiter and Saturn. To record their observations, the Sumerians divided the sky into small groups of stars, or constellations, that they named after animals or objects they resembled. To measure the exact positions of the stars and planets night by night they devised a scale that divided the sky's circular form into 360 degrees. This number was easy to manipulate arithmetically and conveniently close to the number of days in a year. We continue to use this scale to describe circles today.*

The Sumerians noticed that the Sun set in front of a slightly different part of the night sky each night, and returned to the same

---

* 360 is divisible by 1, 2, 3, 4, 5, 6, 8, 9, 10, 12, 15, 18, 20, 24, 30, 36, 40, 45, 60, 72, 90, 120, 180 and 360. The other Sumerian measuring system we still use is for time. A day has twenty-four hours (divisible by 1, 2, 3, 4, 6, 8, 12, 24), an hour has sixty minutes (divisible by 1, 2, 3, 4, 5, 6, 10, 15, 20, 30, 60) and a minute has sixty seconds.

place in exactly a year. Each year it traced the same route through the sky, passing through the same twelve constellations and taking about a month to pass through each. The planets had more erratic movements – sometimes they reversed direction for a while, or disappeared altogether – but they travelled within the same band of sky as the Sun. The constellations along this celestial superhighway included a bull, a crab, a lion, a scorpion, twins, a hunter, a goat-fish and scales. If these seem familiar it is because they are indeed the signs of the zodiac and this was their origin, four or five millennia ago. The word 'zodiac' comes from the Greek for 'circle of animals'. Astronomers call this band of sky the ecliptic.

Once the pattern of celestial movements was deciphered, the Sumerians used historical records to compare the various configurations of the sky with good and bad events in the past. This gave them a basis on which to predict future events. The planets were assigned the characteristics of their respective god – strength, love, beauty, wrath, etc. – and these were then associated with people or events depending on their positions in the sky at different times. For millennia since, astrologers have continued to study the positions of the Sun, Moon, stars and planets for clues to and explanations for earthly events and human behaviour, right up to Mystic Meg.

The Babylonians refined the Sumerian observations and applied mathematics to them. They discovered how to predict events such as lunar and solar eclipses, and the exact timing of planetary movements across the sky. When the Macedonian general Alexander the Great conquered Mesopotamia in 331 BCE, he inherited and appropriated more than 1,000 years of accumulated astronomical knowledge.

Greek philosophers looked to the sky as part of their quest to explain the universe, or *cosmos*. Pythagoras (*c.*570–*c.*495 BCE)*

★ Or one of his followers.

concluded that the Earth was a sphere around which the Moon, planets and Sun rotated, encased in a series of ever larger spheres. Each orb was tuned to a specific musical note that rang out as the planets passed each other, creating a universal symphony he called the 'Harmony of the Spheres'.[12]

Two centuries later Aristotle (*c*.384–*c*.322 BCE) described his version of life, the universe and everything. He proclaimed the spherical Earth sits at the centre of the universe and is comprised of the four earthly elements: earth, water, air and fire. The heavy elements – earth and water – tend down to the centre while the light elements – air and fire – tend upwards and away. Earthly things are imperfect, chaotic and mortal. In the skies above the Earth, the Moon, Sun and planets are made of a fifth element called ether. They are perfect, unchanging and imperishable. Their movement is also perfect, guided by perfect circles and concentric spheres. The fixed stars are in the outermost sphere, beyond which is the spiritual realm, or heaven.

Aristotle's model didn't explain various observed planetary movements – such as Mars appearing to make a loop in the sky from time to time – but nor did it attempt to. In Aristotle's philosophy, the beauty and purity of an argument's reasoning was sufficient proof of its veracity. It was for others to explain any observed inconsistencies.

A few decades after Aristotle, a lesser-known Greek philosopher called Aristarchus (*c*.310–*c*.230 BCE) came up with an alternative explanation for the night sky's movements. His work is long lost, but his contemporary, Archimedes, described Aristarchus' theory:

> that the fixed stars and the Sun remain unmoved, that the Earth revolves about the Sun on the circumference of a circle, the Sun lying in the middle of the orbit, and that the

sphere of fixed stars, situated about the same center as the Sun, is so great that the circle in which he supposes the Earth to revolve bears such a proportion to the distance of the fixed stars as the centre of the sphere bears to its surface.[13]

This Sun-centred (heliocentric) model was dismissed as impossible and soon forgotten. The Earth-centred (geocentric) model remained unchallenged.

For the next few centuries, various Greek mathematicians tried unsuccessfully to reconcile the geometry of Aristotle's model with observed astronomical movements. Finally, Claudius Ptolemy (*c.*100–*c.*170 CE) devised a model that could accurately predict the movements of the Sun, Moon, planets and stars, and was also consistent with Aristotle's cosmos. In *The Almagest*, Ptolemy introduced a new idea to explain the irregular movements of the planets. He proposed that the planets complete little mini-orbits in space called epicycles as they follow their main orbit around the Earth, like an ant on the wheel of a bicycle riding around in a broad circle. Epicycles reconciled observed movements with the geocentric model and could be explained in mathematical terms.

Poland, 1514

With cosmology apparently solved by Ptolemy in the second century, astronomy remained fairly settled in Europe for well over 1,000 years. The Bible included passages that imply the Earth is stationary and the Sun moves,[14] and this was accepted as fact and taught by the Church.

With the European intellectual revival of the twelfth and thirteenth centuries came a revived interest in astronomy, and the Ptolemaic system was taught as part of a basic education.

When the printing press appeared, astronomical texts and tables, calculations and diagrams became relatively widely available.

One beneficiary of these was Nicholas Copernicus (1473–1543), a Polish–Prussian from a well-off merchant background. Copernicus had an extraordinary education financed by a wealthy uncle, first pursuing an arts degree at the University of Krakow, and later studying medicine in Padua and Law in Bologna. He read widely beyond the university curricula, and was particularly interested in the works of the spectacularly named German astronomer Regiomontanus, who established the world's first scientific printing press and published an updated and abridged version of Ptolemy's *Almagest*.[15] While he was studying in Bologna, Copernicus lived with a former pupil of Regiomontanus who was now an astronomy professor, and assisted him with nightly observations of the sky.[16]

By the age of 30, Copernicus was fluent in Polish, German, Italian, Latin and Greek.[17] He had a doctorate in law and was a qualified physician, and an accomplished astronomer. He joined his uncle, now a prince-bishop, as physician, lawyer, administrator and diplomat.

Copernicus continued making astronomical observations and calculations while working for his uncle. He became frustrated with the Ptolemaic model, writing later that, while it was broadly consistent with the numerical data, its cumbersome requirement for multiple epicycles was neither accurate nor pleasing to the mind. He arrived at a simpler arrangement that explained the planets' movements while retaining the idea of perfect spheres and circular motions.[18]

Copernicus' alternative was to put the Sun at the centre of the universe with Earth and the planets orbiting it in ever larger concentric spheres. As he described it, Earth spun on an axis, explaining night and day, and orbited the Sun annually, explaining the seasons, in a sphere situated between Venus and Mars. The Moon was the only body that orbits Earth. With this model he could explain 'the entire structure of the universe and the entire ballet of the planets'.[19] His model still required some epicycles to fit the data, but it was simpler and less convoluted to work with than Ptolemy's.

It is still a matter of scholarly debate whether Copernicus was influenced by, or even aware of, Aristarchus' earlier heliocentric thesis.[20] The Archimedes text wasn't available in print until after Copernicus' death, so most scholars believe he couldn't have seen it. On the other hand, Regiomontanus was a renowned collector of manuscripts, so he may have come across Archimedes. If he had, it would be likely that Regiomontanus had discussed it with his pupil, Copernicus' Bologna landlord.[21]

Copernicus wrote up a brief version of his theory and circulated it among friends and colleagues in 1514. He made no attempt to publish it, however, for fear of ridicule. He knew his theory was radical – after all, it turned 2,000 years of received wisdom on its head and poked a large hole in Aristotle's model of the world. Aristotle's worldview was the linchpin of European natural philosophy in the sixteenth century. Unravelling it would shake the very foundations of Western intellectual culture, threatening to bring the walls tumbling down at a time when the world of knowledge seemed finally to be finding a secure footing after emerging from the Middle Ages.

And if challenging Aristotle was imprudent, a Sun-centred universe also appeared to challenge the words of the Scriptures. The Psalms mention several times that the Earth is fixed and cannot be moved, and the Book of Joshua describes Joshua commanding the Sun to stand still in the sky, not the Earth to stop moving.[22] There were harsh penalties for heresy. And in any case, Copernicus was a pious man with no desire to antagonise the Church.

Clockwise from above: When we look at objects in light and shade, our eyes automatically correct the colours they 'see' to allow for surrounding lighting conditions. In this image Square 1 is the identical shade of grey as Square 2, but our eyes see them as light and dark. (Chris Madden/Alamy Stock Photo); The first of our ancestors to walk upright were the *Australopithecines*, such as this female popularly known as 'Lucy' reconstructed from 3.2 million-year-old fossilised bones found in modern-day Ethiopia. Their bodies were adapted for both climbing trees and walking upright on the ground, and they foraged for food such as fruit and berries. (Momotarou2012/ Wikimedia Commons); The world's first visible animal life evolved underwater around 540 million years ago in a surge of evolution called the Cambrian Explosion. Many of the new species had highly developed eyes. Researchers speculate that it was the development of primitive vision that triggered the evolutionary boom that created the animal kingdom. (Merlinus74/iStockphoto)

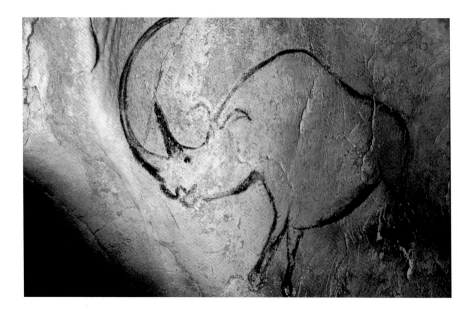

The Chauvet Cave in southern France, discovered in 1994, contains dozens of beautiful images of now-extinct animals common at the time they were painted, 35,000 years ago. They are some of the earliest examples of translating observed reality into images, the first step towards a vision-led culture. See many more images of the Chauvet Cave and other cave art at www.bradshawfoundation.com.

Many of the images of the Chauvet Cave display a graphic quality that is reminiscent of contemporary animations and demonstrates a deep familiarity with image making, though no one knows why the pictures were made deep in a cave with no natural light. (Patrick Aventurier/Getty Images)

Clockwise from above: The Venus of
Willendorf, made in Austria about
32,000 years ago, is large and round
with full breasts and hips, and a
prominent vulva. Her facial features
are absent. This 'Mother Goddess'
portrayal is typical of how females
were depicted in pre-agricultural
communities throughout the world. A
similar figurine, from *c.*6000 BCE, was
found in Çatalhöyük. (MatthiasKabel/
Wikimedia Commons); As
agricultural settlements grew into
cities in ancient Mesopotamia, female
depictions changed dramatically. They
were slimmer and younger looking,
with an emphasis on facial features
that we would today describe as
beautiful, enhanced by cosmetics applied with the aid of mirrors. (Aiwok/Wikimedia
Commons); Scholars tried for years to decipher the mysterious Egyptian writing
system called hieroglyphics. In 1799 Napoleonic soldiers found a stone built into a
wall containing identical inscriptions in Greek, hieroglyphics and Demotic, providing
the key to unlocking hieroglyphics' secrets. The Rosetta Stone is currently in the
British Museum. (Granger Historical Picture Archive/Alamy Stock Photo)

Eighteenth century orientalists used nameplates called cartouches, usually containing royal names, to decipher hieroglyphics. They discovered that hieroglyphics were phonetic – that is, they represented the sounds of spoken language – and not symbolic as had long been assumed. This discovery overturned centuries of erroneous interpretations of Egyptian inscriptions. (Ad Meskens/Wikimedia Commons)

The earliest depiction of spectacles is a 1352 portrait of Dominican monk Cardinal Hugh of St Cher. It appears in a fresco in Treviso, Italy, some 200 miles away from Pisa, where spectacles were invented by an unknown artisan around 1275. The invention coincided with a resurgence of interest in optics in the thirteenth century. (Risorto Celebrano/Wikimedia Commons)

Around 800 CE the Emperor Charlemagne tasked a monk called Alcuin with improving literacy throughout his empire. Alcuin created a new script called Carolingian Miniscule with simply rounded letters and an important innovation: spaces between words. This allowed reading to become a silent, individual and private activity for the first time. (Lebrecht Music & Arts/Alamy Stock Photo)

In his eleventh-century treatise on Optics the Arabic scholar Alhazan described a natural optical phenomenon that had been observed for centuries. Light passing through a small hole into a darkened space can project an image of the scene outside onto a light-coloured surface. This is known as a *camera obscura*. The phenomenon also explains how images form on the eye's retina. (Mario Bettini, 1642)

Giotto's Scrovegni Chapel frescoes, painted in 1300, utilised naturalistic painting techniques not seen in Europe since Roman times. Some scholars believe Giotto was influenced by the enormous intellectual interest in optics in the thirteenth century, prompted by the translation of Alhazan's *Book of Optics* into Latin.

Robert Hooke's 1665 *Micrographia* illustrated the curious structures that appeared in fine slices of cork when viewed under a microscope. He called them 'cells' after the small enclosures where monks copied manuscripts. The book inspired Antoni van Leeuwenhoek, who went on to discover bacteria and spermatozoa, though it was another two centuries before Pasteur developed his germ theory. (Hooke, 1665/Wellcome Collection)

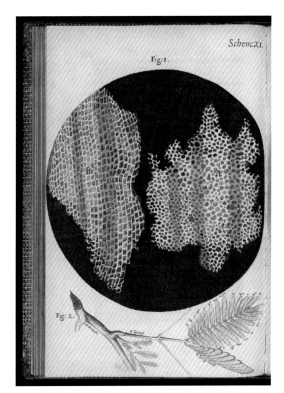

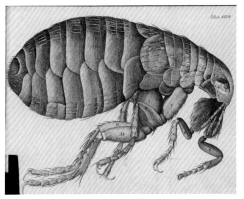

Above left: Micrographia contained dozens of illustrations of details never seen by the naked eye. It inflamed the public's imagination and transformed the understanding of nature and matter. The entire book can be viewed online at archive.org. (Hooke, 1665/Wellcome collection)

Above right: Sir Isaac Newton invented the reflecting telescope – using mirrors rather than lenses – in 1668. In 1781, William Herschel used an enormous reflecting telescope to discover the first new planet found since antiquity. His discovery of Uranus orbiting outside Saturn expanded the size of the known universe fourfold. (Wellcome Collection)

A PEEP AT THE GAS LIGHTS IN PALL-MALL.

Illumination by gaslight made night-time city streets safe for the first time. The streets of London filled with people admiring the revolutionary light with no apparent source. A cartoon of the time satirised the varied reactions, from wonder at the technology to concern about the potential loss of 'night-time trades'.

The period from the sixteenth to the late eighteenth centuries spans the movements known as the Scientific Revolution and the Enlightenment. Both prompted a fervent interest in visual observation and in capturing and cataloguing images of nature. The *camera lucida*, invented in 1806 by William Wollaston as an aid to drawing, is a prism on a small stand. It reflects the image being observed over the sheet of paper, allowing the user to trace it.

John Constable (1776–1837) and his contemporary J.M.W.Turner (1775–1851) were the first modern artists to embrace the depiction of nature and landscape in all their humble details as subjects worthy of artistic attention. Their work inspired later artistic innovators. (The Picture Art Collection/Alamy Stock Photo)

The first permanent photograph ever taken was in 1826 by inventor Joseph Nicephore Niépce. It was the view from the window at his house in Le Gras, France. Niépce died a few years later but his partner Louis Daguerre took his work forward and unveiled the world's first commercially viable photography process, the eponymous daguerreotype, in Paris in January 1839.

Daguerreotypes were an instant sensation for a reason Daguerre himself had failed to anticipate: everyone from farmhand to nobleman wanted to have their picture taken. Portraits were the 'killer app' for daguerreotypes. (Gift of Samuel J. Wagstaff, Jr/The J. Paul Getty Museum)

Early daguerreotypes of street scenes appear uninhabited because moving elements like people, horses and carts didn't register in the long exposure times required to capture an image. This picture captured the first photographed humans by accident: a shoeblack and his client, relatively still among the bustle of Paris's Boulevard de Temple.

When news of Daguerreotype's process reached Philadelphia, local chemist Robert Cornelius immediately set about making himself a camera. He took it outside and stood in front of it in the bright sunlight for long enough to make the first known photographic portrait, and the first ever selfie. (Library of Congress)

In 1860 an unknown political candidate had his portrait taken in New York by photographer Matthew Brady. When the candidate, Abraham Lincoln, was elected sixteenth President of the United States he credited Brady's iconic portrait for 'making' him so. (Library of Congress)

Escaped slave Frederick Douglass became a famous orator and writer and a leader of the abolitionist movement. He used photographic portraits of himself to change public attitudes towards black people. Handsome, well groomed, wearing elegant clothes and with dignified comportment, he demonstrated that black people were just as human and worthy of admiration as whites. Douglass was the most photographed American of the nineteenth century. (National Portrait Gallery, Smithsonian Institution)

Questions about truth and photography are as old as the medium. The photographer of this 1863 image of the Battle of Gettysburg known as *A Harvest of Death*, Timothy H. O'Sullivan, has been accused of altering scenes to increase their emotional impact. Does moving a gun render an image untrue? (Library of Congress)

Scientists were interested in photography that showed details invisible to the human eye. Eadweard Muybridge set up banks of cameras to capture the details of human and animal movement, such as this 1881 image of a man jumping a hurdle. His work changed the understanding of movement and anatomy. (Library of Congress)

The photographic pioneer William Fox Talbot – inventor of the first positive-negative photographic process – experimented with the artistic possibilities of photography. This 1844 Talbot photograph, *The Open Door*, explores the idea that what is missing from an image can be more interesting than what is shown. (Gilman Collection, Purchase, Joseph M. Cohen and Robert Rosenkranz Gifts, 2005/The Metropolitan Museum of Art)

Some early attempts at artistic photography embraced classical themes, such as this 1857 image by Oscar Gustave Rejlander, an allegory called *The Two Ways of Life*. As a montage composed from more than thirty negatives, it is a technical masterpiece but the style and subject look somewhat uncomfortable as a photograph.

Nature photography was (and remains) a highly successful artistic endeavour. Carleton E. Watkins captured the grandeur of California's Yosemite using enormous cameras, including this 1865 image. His work prompted President Lincoln to legislate to protect Yosemite and paved the way for the US National Parks Service. (The J. Paul Getty Museum)

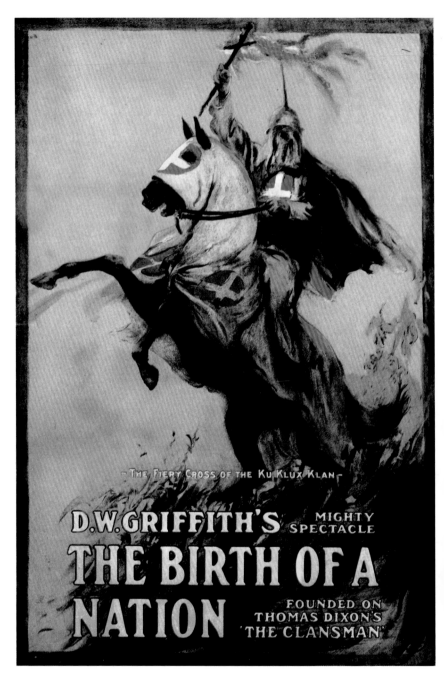

As motion pictures became popular entertainments, filmmakers' ambitions grew. D.W. Griffiths was the first to employ sophisticated film editing to enhance realism and emotional impact, to incredibly powerful effect. His 1915 film *Birth of a Nation* caused a real–life resurgence of the Ku Klux Klan after decades of oblivion. (Pictorial Press Ltd/Alamy Stock Photo)

What was more, while it may have satisfied the arcane concerns of a few astronomers, heliocentrism blatantly offended the everyday experience of the senses. A spinning 'whirligig' was used as an instrument of torture at the time: would God create a similarly spinning planet for his beloved Man? Besides, to complete a full rotation in a single day the Earth would have to spin at 1,000 miles an hour. That was simply inconceivable! Even if it were possible, wouldn't people be permanently nauseous? Where was the whoosh and the wind of such a movement? Why did everything on the Earth's surface not simply fly off, as an unsecured barrel rolls off a moving cart? Above all, perhaps, what would it say about mankind's place in the universe if, instead of sitting at its centre, Earth was just another vagabond star wending its way around the Sun as one among many?

In sum, Aristotle's cosmos had a natural logic that accorded with most people's daily experience: there was very little incentive to challenge it.

Nevertheless, while Copernicus carried on with his professional duties over the subsequent years, he quietly set about gathering the evidence and mastering the arguments and calculations to underpin his radical proposal. Meanwhile, as the Reformation took hold across parts of Europe, news of Copernicus' theory slowly made its way around scholarly circles. In 1533 the humanist Pope Clement VII heard of the theory and asked to have it explained to him, and in 1536 a senior cardinal wrote to Copernicus expressing his admiration and asking for details. Martin Luther heard of the idea but was considerably less encouraging. The once 'upstart priest' was by now a powerful man, and he reputedly dismissed Copernicus as a fool.[23]

Copernicus, a resolute Catholic, was finally persuaded to publish by a young Protestant mathematician called Osiander, who helped him assemble his work and supervised its printing. *On the Revolutions of the Planets* (*De Revolutionibus Orbium Coelestium*) was published in 1543, thirty years after Copernicus had originally outlined his heliocentric theory. By this time Copernicus was old and ill. As legend has it, he received the first printed copy of his life's work on his deathbed, and promptly expired.[24]

A Very Quiet Revolution

Considering the potentially revolutionary implications of Copernicus's book, its initial impact was inconsequential, though it was a highly technical book so its audience was admittedly limited. Critics made arguments against it along the practical lines described above, and there were some theological objections to a moving Earth and stationary Sun. However, Copernicus was not denounced, *De Revolutionibus* remained in print and a second edition was produced twenty years later.[25]

What may have saved *De Revolutionibus* from further censure was an anonymous preface – added at the last minute without Copernicus's knowledge by his pupil, Osiander – that said the heliocentric theory proposed could be interpreted as a mathematical rather than a physical model. The model was mathematically consistent with the observed astronomical phenomena, it said, but it need not be true nor even probable as long as the mathematics worked.[26] Scholars have debated the purpose of this preface, but it may well have kept the implications of the book sufficiently ambiguous to keep it out of the hands of the Roman Inquisition.

Or perhaps it wasn't overly threatening because it just wasn't convincing enough?[27] Apart from its apparent common-sense flaws there were still mathematical problems with the theory. A couple of decades after *De Revolutionibus* was published, a Danish astronomer called Tycho Brahe set up the first observatory in Europe and spent years making hundreds of observations using the most advanced measuring instruments. With access to the most comprehensive data ever available, Brahe rejected the Copernican model and came up his own modified version of the Ptolemaic cosmos.

By the turn of the seventeenth century, astronomers could choose from three possible models of the universe: Ptolemy, Copernicus or Brahe. None was definitive. As Osiander had said in his unauthorised preface, it was impossible to know for certain how the planets actually moved; therefore it made sense to follow whichever assumptions gave the best mathematical result in the simplest manner. In the absence of further evidence, the debate was unresolvable.

The Hague, 1608

In September 1608, a Dutch spectacle-maker from Middleburg travelled to The Hague carrying a letter from local officials recommending him to Prince Maurice, the commander-in-chief of the Dutch forces. Maurice was conducting a peace conference with representatives of the Dutch Republic's long-term enemies, Spain and France. The letter informed the prince that the bearer, Hans Lipperhey, claimed to have invented an apparatus for 'seeing remote objects as if they were near'. The device was an enclosed tube with a larger convex lens (the objective) at one end and a smaller concave lens at the other. If it worked as promised it could be a valuable military instrument.

The prince was impressed, and proudly demonstrated the discovery to the French and Spanish envoys. He paid Lipperhey a generous sum to produce several more telescopes exclusively for him. A week later, Lipperhey submitted a patent application. The patent office was less forthcoming than the prince and refused Lipperhey a patent on the grounds that the device was too easily copied to be patentable.

Unfortunately for Lipperhey, the patent office was proved correct almost immediately. Tales of the telescope spread around Europe, as the Spanish and French commanders promptly reported back to their respective courts.[28] The ruler of the Spanish Netherlands got hold of an instrument and managed to copy it. He sent one to a cardinal in Rome,[29] while the French ambassador dispatched a soldier to Paris with instructions on how to copy the device for the king there. Claims of prior invention emerged and carried on for years.[30]

Others may well have come up with the same idea before Lipperhey – it was after all a fairly simple one and could have been arrived at by accident while playing around with lenses many times over. But it is now generally agreed that, in taking it into the public arena, Lipperhey was the first to recognise the telescope's potential as something more than a curiosity, even if he wasn't the first person to make one.

In any case, beyond the honour and glory of being the telescope's inventor, by late 1608 it was obvious there was no hope of Lipperhey gaining any exclusivity or privilege in making it. Telescopes could be bought in Paris by April the following year, in Milan by May, and Venice and Naples by the summer.[31]

Notwithstanding this flurry of interest, many people found the first telescopes disappointing to look through. They were made with spectacle lenses and only magnified about three times. Their field of vision was very narrow, making them difficult to use. The images were considerably worse than you would see through the cheapest telescope available today. They fell far short of the possibilities that had been dreamt of by optics scholars for centuries. Four hundred years earlier, Roger Bacon speculated on the possibilities of lenses to 'cause the sun, moon and stars in appearance to descend here below' and allow men to 'read the smallest letters and number grains of dust and sand' from a great distance.[32] Lack of sufficiently high-quality glass had prevented his ideas from becoming a reality for all this time, and it was only marginally possible now. When the famous scholar Giambattista della Porta first saw a telescope in 1609 he was so unimpressed he told a friend he thought the whole thing was a hoax.[33]

A brilliant Florentine mathematics professor soon changed that, however. By 1609, Galileo Galilei (1564–1642) had already invented several scientific instruments and conducted various experiments to challenge received scientific theories. He was a bon viveur and always short of money, so he supplemented his professorial salary by selling instruments and treatises to his richer students.[34] When he heard of the invention of the telescope, he set out to make one for himself, grinding his own lenses more and more precisely. After several months, he succeeded in making a telescope that made objects appear 'thirty times closer and a thousand times larger'[35] than through natural vision. While this was an exaggerated claim, his telescope was a huge improvement on the Lipperhey model.*

Galileo trained his far superior spyglass on the night sky and saw things no one had ever seen before. He studied the Moon first and discovered it wasn't smooth, uniform and precisely spherical as Aristotle had assumed. Rather, it was uneven, rough and covered with cavities and mountains, not unlike Earth.

---

* Galileo measured an instrument's power by looking simultaneously at two circles of different sizes, viewing one with the naked eye and one through the telescope. When the two circles appeared the same, the telescope's power was equal to the ratio of the circles' sizes.

Turning his attention to the sky beyond, he observed that the planets looked round 'like little Moons' through the telescope, but the fixed stars continued to look the same as with the naked eye, only brighter. His telescope revealed a host of other stars 'so numerous as almost to surpass belief'.[36] Within the constellation of Orion, for example, he counted another 500 stars, while the famous Pleiades constellation contained at least forty others, densely grouped. He saw that the Milky Way, long the subject of disputes as to its true nature, was a mass of innumerable stars grouped together in clusters, and that the nebulous stars were groups of small stars arranged in wonderful ways.

The most important discovery of all was four new wandering stars in the vicinity of Jupiter. He didn't realise they were planets at first, but they caught his attention one night because they formed a straight line with Jupiter, parallel to the zodiac band. The following night he looked at them again and saw that they were still in a straight line but their order had changed (what powers of observation he must have had to distinguish between them!). When he observed them a third time they had changed places yet again. These observations continued for the next two months, as he watched the four new planets follow Jupiter slowly across the sky, changing places every night but staying within the same distance of the larger planet, and always in a straight line. Galileo concluded that these planets must be moons orbiting Jupiter in the same way our own Moon orbits Earth. This was a momentous discovery. In honour of – or rather blatant sycophancy to – the Grand Duke of Florence, Cosimo de' Medici II, he named the new planets the Medicean stars.

Galileo published these revelations in March 1610 in a short, illustrated book entitled *The Starry Messenger*. He pointed out that the moons of Jupiter provided a 'fine and elegant argument' for the Copernican system, as they proved that not all celestial bodies revolved around Earth. Galileo promised to present in due course a system that would prove that the Earth was a wandering body circling the Sun, just as Copernicus had said.

The book was an international sensation and, as he had hoped, his revelations caught the attention of the Grand Duke Cosimo, who appointed him mathematician and philosopher to his court in Florence. Now Galileo had time and resources to pursue astronomical observations and experiments, and make ever better telescopes. Over the next two years he discovered that Venus goes through phases much like the Moon, indicating that the planet orbits the Sun rather than the Earth, and that the Sun has sunspots, providing another challenge to the long-held view of the 'perfection' of the celestial bodies.

Galileo's observations provided mounting evidence for a Sun-centred universe. But many of his contemporaries remained steadfastly sceptical and some suggested that images seen through a telescope were not to be trusted, at least as far as observations of the sky were concerned.[37] A few even refused Galileo's offer to look through the device themselves, much to his frustration. More dangerously for Galileo, his pugnacious manner and intellectual aggressiveness was attracting enemies.

For more than seventy years, the Catholic Church had tolerated Copernican ideas, choosing to accept them as a mathematical rather than physical model. Galileo's observations and accompanying comments challenged this status quo, as they suggested that the heliocentric model may, in fact, represent physical reality. This was bound to be controversial. From 1613 rumblings began that the idea of a spinning Earth was against Scripture.[38] In response to these rumblings, Galileo wrote a long letter to Duke Cosimo's mother, the Dowager Grand Duchess, defending his views. He dismissed those who challenged his view on theological grounds in the most disparaging terms, quoting St Augustine, who said theologians should concern themselves with matters of faith rather than questions of nature, and also a contemporary cardinal who had said, 'the intention of the Holy Spirit is to teach us how one goes to heaven, not how the heavens go'.[39] This further alienated his critics.

In 1616 Galileo ended up before the Roman Inquisition, who condemned heliocentrism as foolish and absurd, and ordered Galileo to abandon the Copernican theory, although Galileo himself was not rebuked.[40] Copernicus's *De Revolutionibus* was placed on the Index of Forbidden books, along with any other books with a similar view.[41]

In 1620, ten years after *The Starry Messenger* was published, the English politician Sir Francis Bacon published *Novum Organum*, in which he proposed that empirical evidence and experiment – attained systematically, he said, with the aid of newly invented instruments, such as the telescope, to augment the limited human senses – should overthrow the traditional model of knowledge attained through logic and reasoning and hasty generalisations. Three years later, Galileo published his own essay on the nature of science and scientific enquiry. It contained one of the most famous statements in the history of science:

The Universe cannot be understood unless one first learns to comprehend the language and interpret the characters in which it is written. It is written in the language of mathematics, and its characters are triangles, circles, and other geometrical figures.[42]

When Rene Descartes published his thesis on Cartesian geometry a decade later, he gave the world a method to translate time and space into mathematical terms, just as Galileo had described.[43] Sir Isaac Newton, born the year Galileo died, developed laws of motion and gravity that explained the movement of terrestrial objects and the solar system in terms of mathematical relationships. Newton's laws not only provided a compelling explanation of the Copernican cosmos, they were the ultimate demonstration of Galileo's assertion that the universe was written in the language of mathematics. They also provided a new holistic explanation of the universe, striking the final blow to Aristotelean philosophy.

Florence, 1632

Pope Urban VIII received Galileo's 1622 philosophical essay warmly, and Galileo was emboldened to return to his Copernican studies. In 1632 Galileo published *A Dialogue Concerning the Two World Systems*, his comprehensive case for the Copernican system. It was written in the classical scholastic form of a conversation between three friends, in which the pro-Ptolemy (geocentric) character Simplicio was portrayed as a bumbling imbecile. Galileo's enemies claimed Simplicio was modelled on the Pope, and the unflattering portrayal infuriated the pontiff. The book was immediately referred to the Roman Inquisition and once again Galileo was summoned to Rome to face trial. Galileo, by now old and suffering from illness, capitulated. He declared he had been mistaken and was forced to kneel before the Inquisition and recant his belief in the Copernican system.[44] He returned to Florence and spent the next nine years under virtual house arrest at his villa until his death in 1642.

The Terrestrial Telescope

The usefulness of the telescope wasn't confined to starry-eyed astronomers. Spyglasses quickly became an indispensable tool on land and, especially, at sea. In the early 1600s, the Age of Discovery was drawing to a close and European powers were racing to colonise global empires. This demanded huge maritime fleets. Sea travel, propelled by sail, was at its apex throughout this century and the next. No self-respecting sea captain would be seen without a spyglass, as any portrait of a mariner of the period will attest.

Lesser mortals found telescopes useful in daily life, although not always for reputable purposes. The diarist Samuel Pepys commented in 1667: 'I did entertain myself with my perspective glass up and down the church, by which I had the great pleasure of seeing and gazing at a great many very fine women.'[45]

In the next century the opera glass appeared, a small personal telescope initially with a single tube (monocular) and later in binocular form. These were often made of precious materials and exquisitely decorated, and they became an essential accessory in fashionable society.

Hot on the heels of the telescope came the earliest microscopes, probably also invented by Dutch spectacle-makers.[46] Like the first telescopes, early microscopes were not particularly effective, magnifying only around nine times and giving a blurry image. At first they were of more novelty than scientific value, and were used to look at fleas and other tiny insects, earning them the name 'flea glasses'. Interest in microscopes took off in 1665 when Robert Hooke published *Micrographia*, a magnificent book filled with stunning illustrations of microscopic details including a gigantic flea and microscopic fungi growing on leather.[47] Hooke introduced the term 'cell', derived from the monks' cells in scriptoria, to describe the patterns he saw when observing pieces of cork through the microscope. The book inspired Antoni van Leeuwenhoek, who became the first person to observe bacteria, single-celled protozoa and spermatozoa.[48] Despite these significant visual breakthroughs, it was another 200 years before Louis Pasteur developed his germ theory and made the link between microorganisms and disease.

Seeing Beyond Galileo

Galileo's trial and conviction shut down one individual, but it was too late. Pandora's scientific box had been opened, and there was nothing the Church could do to prevent the ongoing investigation of the skies, nor the ongoing reassessment of the Scriptures' literal truth. Not that the debate was purely one between science and religion, as is often portrayed. At the time and for many years after there were scientists who weren't convinced by the Copernican/Galilean cosmos, and churchmen who were.[49] What Galileo's early telescopic observations, his writings and his subsequent trial had unleashed was a search for truth and, increasingly, the belief that truth was to be found through evidence obtained by observation, experimentation and calculation, rather than doctrine.

Useful and amusing as it was in terrestrial applications, it was the telescope's use in astronomy that changed seeing forever. By providing evidence for the Copernican cosmology, the telescope upended the received view of mankind as the centre and indeed the very point of the universe. This was a very unsettling idea in itself, but it also called into question the literal interpretation of the Bible. If passages referring to a stationary Earth were not correct, what other parts of the Scriptures might be wrong? Europe was still in the grip of religious turmoil in the wake of the Reformation and Counter-Reformation of the previous century. Such disruptive ideas must have been extremely unwelcome.

The notion of the Moon and planets as tangible, solid places rather than ethereal skylights stirred musings on the possibility of life beyond Earth. The playwright Ben Jonson explored the idea of the Moon as another New World in a 1620 stage production that featured mute and naked Moon dwellers.[50] In 1638 a clergyman called John Wilkins, later brother-in-law of Oliver Cromwell and a founder of the Royal Society, asked 'if our earth were one of the Planets ... then why may not another of the Planets be an earth?'[51] and went on to consider potential methods of travelling to the Moon. The text of Hooke's *Micrographia* wondered aloud whether future inventions might allow us to discover living creatures on the Moon or other planets.[52]

As well as radically altering the way people saw themselves within the universe, the telescope – aided by Galileo's erudite pronouncements on science and observation – prompted a radical shift of worldview in the field of knowledge. Where Copernicus's theory had dealt a glancing blow to Aristotle's system of thought, the wound the telescope inflicted was fatal. The telescope was the first scientific instrument to reveal what human observation alone could not. It demonstrated that the natural world was more complex than Aristotle's beautiful hypothesis supposed. The heavens were not perfect and unchanging; the Earth was not stationary. Crucially, Galileo and his contemporaries recognised that to truly understand nature they had to make careful and repeated observations rather than rely on logic and reasoning.

An Ever-Expanding Universe

The final years of Galileo's life took on a further element of tragedy when he lost his eyesight. He wrote to a friend in 1638:

> The universe which I with my astonishing observations and
> clear demonstrations had enlarged a hundred, nay, a thousand
> fold beyond the limits commonly seen by wise men of all
> centuries past, is now for me so diminished and reduced, it
> has shrunk to the meager confines of my body.[53]

What he couldn't know was that his initial discoveries were
merely the first in a long line of firsts in the expansion of
the universe.

Galileo's early monopoly on the manufacture of superior
telescopes was brief, but his priority in terms of discoveries lasted considerably longer. It was nearly fifty years
before Christian Huygens discerned the rings of Saturn
and observed its moon, Titan. Then in 1781 a German-born
self-taught astronomer discovered a new planet, the first
such discovery since antiquity. William Herschel was using
a reflecting telescope, first developed by Newton in 1668,
that he had built from scratch with his brother and sister
in the back garden of their home in the spa town of Bath,
England. The discovery of the new planet, eventually named
Uranus after the Greek god of the sky, expanded the size of
the known universe fourfold.[54]

Later astronomers observed that Uranus's orbit was slightly
but systematically different from what it should have been
according to Newton's laws of gravity. They deduced that
an unknown planet must be exerting a gravitational pull on
Uranus and thereby disrupting its orbit. A French astronomer,
Le Verrier, calculated where this unknown planet should be and
sent his predictions to the Berlin Observatory in September
1846. That very night, a new planet, Neptune, was observed for
the first time almost exactly where Le Verrier had said it would
be.[55] Neptune was orbiting about another billion miles beyond
Uranus. The solar system expanded again.[56]

After his success with Uranus, William Herschel built another telescope so large that people could walk through it. Working with his sister Caroline, he observed stars that moved in relation to one another, proving Newton's law of gravity to be truly universal. The Herschels surmised that the Milky Way was a roughly disc-shaped system of around 300 million stars in a grouping later called a galaxy.[57] This enormous universe was unimaginable degrees larger than the cosmos had been believed to be 200 years earlier. But of course, much more was to come.

In 1838, another self-taught German astronomer called Friedrich Bessel discovered the final piece of evidence that proved the Copernican cosmos beyond doubt when he detected something astronomers had been seeking for two centuries. Stellar parallax is the shift in relative position of stars seen from different distances, like seeing objects at different distances from a travelling car: closer objects pass more quickly than those further away. His discoveries proved beyond doubt that the Earth moves. They also allowed him to calculate the distance to the nearest star, which he estimated at about 65 trillion (10^{12}) miles from Earth, more than five times previous estimates.[58]

At this point astronomers needed a new language to describe distances. Talking in terms of millions or billions of miles just about worked within the solar system, but as they were discovering it was completely unwieldy when describing a galaxy. The concept of the light year – the distance light travels in one year – was gradually adopted. One light year is a little under 6 trillion miles. Earth to Neptune is about 4 light hours. The nearest star to Earth, Proxima Centauri, is 4.22 light years away. Incidentally, modern astronomers now use another new term, the parsec, which can be measured more precisely and equates to about 3.26 light years.

All these expansions of the known universe were nothing, however, compared with the observation by Edwin Hubble in 1923 of Star V1, a pulsating star within the Andromeda nebula. Observing V1 from the Mount Wilson observatory in Los Angeles, Hubble calculated that V1 is almost a *million* light years away. This placed it outside our own Milky Way galaxy, which is only 100,000 light years across, and led to another astounding conclusion.

The Milky Way is not unique. Not only not unique, but one of *millions* of galaxies. And not only that, Hubble's observations provided the first evidence that the universe is not stable but expanding.[59]

The universe, as we thought we knew it, was not just a solar system, nor even an enormous galaxy, but an unknowable space within which our planet – once believed to be the centre of that universe – spins alone, anonymous and invisible.

In the last decade of the twentieth century, NASA launched four giant telescopes into space, the best known of which is the Hubble Space Telescope. These have continued to challenge our notion of space and our place in it. In the 1990s astronomers believed there were around 200 billion galaxies in the universe. In 2016 Hubble data suggested this was an underestimate by a factor of ten.

The Eye, Extended

The telescope was the first invention to extend a human sense, taking seeing far beyond anyone's natural ability and beyond the bounds of our earthly home.

In 400 years our understanding of mankind's place in the world has been comprehensively transformed by observations made through ever more powerful telescopes. We have learned that the Copernican model of the solar system was broadly correct but it was only the beginning of a much longer story. It is almost unimaginable that people could ever have believed that Earth sat stationary at the centre of the universe, attended by our very own Sun, Moon and planets, and comforted nightly by the twinkly canopy of fixed stars. But equally difficult to imagine is that Earth is but a tiny planet orbiting a middle-aged star, one of around 300 billion stars, sitting on the arm of a fairly ordinary, spiral-shaped galaxy, in turn one of 2 *trillion* galaxies within an ever-expanding universe.

From our position at the centre of our own universe, we have learned that we are smaller than a grain of sand in a giant desert. And despite everything we've learned in 400 years, the stars remain enigmatic, forever beyond our reach.

As telescopes become more and more powerful they see deeper and deeper into history. The images we see are not windows into the worlds of our distant neighbours but snapshots of a history that unfolded eons before our species even existed. Modern telescopes see stars and galaxies being born and dying, re-forming and collapsing. Telescopes have observed galaxies up to 13.4 billion light years away, which equates to roughly 13.4 billion years *ago*. That is, astronomers are getting closer and closer to observing galaxies as they were just after the Big Bang. They fully expect the next generation space telescope, the James Webb due to launch in 2021, to observe the formation of the first galaxies. The thought of astronomers seeing the beginning of time itself begs the question: what comes after – or rather before – that?

If all that feels overwhelming, consider further that physicists estimate that what we can see – and will ever see, in the sense of detecting light – accounts for only around 5 per cent of the universe's total mass. The other 95 per cent of the universe is accounted for either by dark matter (27 per cent) or dark energy (68 per cent). Whatever these substances are (and no one knows), they are invisible to existing man-made instruments.

Here we come full circle with Ptolemy, Copernicus, Brahe and Galileo. Like their predecessors, modern physicists such as the late Stephen Hawking debate potential cosmologies based on the information they have available and sensible guesses about what they believe may be discovered in future. But until new modes of observation are invented, no single theory can be proved conclusively correct – just as it took many years for the Copernican cosmos to be proved beyond doubt.

Alone in infinite space, we still cling to the hope of finding life beyond Earth. Modern telescopes have found a handful of Earth-sized planets orbiting other stars in what may be a habitable zone: the so-called Goldilocks Zone, where water would neither freeze nor vaporise. The hope is that future telescopes might show signs of life.[60] Closer to home, the search for extraterrestrial life has moved from visual to physical. NASA has four active rovers currently exploring Mars and analysing gases and solids on and below the surface. In June 2018 the *Curiosity* rover discovered possible organic material within rocks and variable levels of methane emissions that suggest the possibility of life on the planet at some point in its history.[61]

And what of the 95 per cent of the universe resolutely invisible to electromagnetic detection? Theoretical physicists trying to understand dark matter are using a giant earthbound detector called the LIGO.* This is not a telescope but a machine that 'listens' for gravitational waves from space. With this tool, researchers hope to detect and eventually, perhaps, explain dark matter.

---

* Laser Interferometer Gravitational-Wave Observatory. This was the experiment that detected the first gravitational waves.

Wonder and awe still abound, then. And the greatest and most awesome wonder of all is that our insignificant species, marooned on our tiny blue orb in the middle of infinity, has managed – through curiosity, ingenuity, invention and imagination – to reach out into that vast, unfathomable space and see what lies beyond.

IN LOVE WITH NIGHT:
INDUSTRIALISED LIGHT

All the world will be in love with night
And pay no worship to the garish sun.

William Shakespeare (1585–1613), *Romeo and Juliet*, 1595

Firelight, Domesticated

For hundreds of thousands of years after humans tamed it, firelight remained our only source of artificial illumination. Firebrands became torches very early in human history when people started wrapping sticks in skins soaked in pine pitch or animal fat, thus burning the fuel rather than the wooden stick underneath.* Their sensual, slightly menacing flicker has lit human journeys for thousands of years, and still do.

While people were still living outside in nomadic, hunter-gatherer groups, a tamer form of firelight appeared, smaller and

---

* *Torch* comes from the Latin *torquent*, meaning twist.

calmer than the torch and giving a relatively long-lasting, safe and low-maintenance light. This domesticated lighting technology was the wick: a twisted bundle of absorbent fibre that draws up liquid fuel in a capillary action. When lit the fuel vaporises around the wick and the gases burn, giving off heat and light from a characteristic teardrop-shaped flame. The first wicks were made of twisted strands of moss or juniper and used in makeshift stone lamps fuelled by animal fat. Dozens have been found in south-west France; seventy were found stored together at the Lascaux Cave.[1] They produce less light than a modern candle but enough to guide a person through a cave or illuminate fine work.

Clay lamps came next, appearing around the sixteenth century BCE and giving about as much light as one or two modern candles.[2] They became ubiquitous around the Mediterranean, where olive oil was plentiful, and by Roman times were being mass-produced. Many had a maker's mark stamped on the base, making them perhaps the first example in history of a branded household appliance.

The Egyptians used dried rushes dipped in fat to make lights and the Romans adapted these to make the first European candles (although the Chinese, Japanese and Indians also invented candles independently at different times).[3] A candlewick is similar to a lampwick but in the case of a candle its solid body *is* the fuel. When lit, the flame melts a small amount of fuel at the base of the wick/top of the candle, and this is drawn up into the wick and vaporised. As the vaporised gas burns, it consumes the top part of the wick and melts a little more of the fuel, and so on.

Roman chandlers made candles by dipping a wick repeatedly into melted tallow (rendered animal fat). This has been mechanised and the fuel has changed over the centuries, but essentially this process is still used today. In Northern Europe, where olive oil for lamps was scarce, tallow candles were the main source of artificial light for centuries, despite their rather weak light and unpleasant smell. In rural areas, candles were made at home with boiled animal fats – a smelly and messy process that was one of the least favourite tasks of the country housewife.[4]

In cities, chandlery was an important trade. The London Tallow Chandlers Company was formed in 1300 and became one of the City of London's Worshipful Livery Companies in 1462. The City of London ranked the Tallow Chandlers' and Wax Chandlers' Companies twentieth and twenty-first in order of importance, outside the Great Twelve* but above butchers, plumbers and carpenters, and immediately below bakers.[5]

Beeswax candles came into use in the Middle Ages but were prohibitively expensive for ordinary people: they were the preserve of royalty and the Church. In the eighteenth and nineteenth centuries it was discovered that spermaceti from the head cavity of sperm whales made brilliantly white, clean-burning candles, and the whaling boom that followed nearly wiped out the sperm whale. They were saved from extinction just in time in the late nineteenth century when paraffin wax derived from petroleum displaced spermaceti. Paraffin was the first fossil fuel. It was cheap, burned clean and bright, and is still the main component of mass-produced candles today.

As their longevity attests, candles are a remarkable technology. Portable and packable, they are self-contained and require no fuel beyond an igniting light. They could be made at home with the by-products of everyday life, and consumed themselves, leaving little mess. With a suitable lantern to protect its flame, a candle could travel outside. Its characteristic flame is generally steady and safe, kept constant by the laws of physics,

---

* The Great Twelve were, in order, Mercers, Grocers, Drapers, Fishmongers, Goldsmiths, Merchant Taylors, Skinners, Haberdashers, Salters, Ironmongers, Vintners, and Clothworkers.

and those desirous of more light could multiply its luminosity again and again limited only by money and space. Where these were abundant, so was the brilliance: in 1688 the French Sun King, Louis XIV, reportedly used 24,000 candles to light a party at his palace in Versailles.[6]

Pre-Industrial Societies, Night Time

In agricultural societies, the busiest times of year were also the lightest. The hard work of ploughing, planting, tending and harvesting all happened in the warmer, lighter months. Winter was the time to hunker down around the hearth, but where timber was scarce, maintaining a fire much beyond mealtimes was not practical. The alternative was to snuggle up in bed.

Bed-dwellers weren't always sleeping, however. In his history of night-time, Virginia Tech history professor A. Roger Ekirch reveals that in pre-industrial societies it was perfectly normal for people to split their sleep during the night.[7] The typical sleeping pattern was a first sleep of around four hours, an hour or two of wakefulness, then a second sleep of another four hours or so. The wakeful period was known as the watch, and while it would often be spent lying quietly in bed, Ekirch also found references to people doing chores, reading, smoking, praying, lovemaking and even visiting neighbours in the middle of the night. Ekirch cites references to the watch in literature from Homer's *Odyssey* through Chaucer, John Locke and Miguel Cervantes to Charles Dickens in 1840.[8] He quotes medical text books that advised sleepers that their digestion and restfulness would be improved by taking their first sleep lying on their right side and their second on the left, and told couples hoping to conceive to delay their efforts until after the first sleep, when they would do it 'better'.[9]

This pattern of segmented sleep seems to be a natural response to long, dark nights, and continues in small-scale traditional communities. In the 1970s the anthropologist Polly Weissner described a typical night amongst the Ju/Hoansi Bushmen of Namibia:

> As the fire faded to coals, people returned to their respective hearths to settle in for the night as sleepiness set in. Hours later, around 2:00 AM (the 'little day'), some adults awoke, smoked, stoked the fire to deter predators, and chatted for a short time.[10]

In a 1990s experiment at the US National Institute of Mental Health, volunteers were subjected to fourteen hours a day of darkness, with no artificial light. After a few weeks they began to follow the ancient pattern of four hours' sleep, two hours awake and four hours' sleep.[11]

As artificial light was industrialised in the late eighteenth and nineteenth century, references to first and second sleep die out. By around 1920 the idea of a naturally broken sleep was forgotten and replaced by the notion that a 'good' night's rest was sleeping eight hours straight through the night.

Shutting in Time

When people started settling into permanent homes, the campfire lost its place as the centre of tribal life and was replaced by the hearth. It was the place where people gathered to attend to the chores generated by the day's work and to keep company together. Eight-thousand-year-old remains at Çatalhöyük show evidence of people making and repairing tools in front of their hearths.

With people and fires tucked safely indoors, night-time outside became a dreaded place once again. Wild animals were no longer a threat, but there was a new nocturnal enemy: other people.

As settlements grew into villages, towns and eventually cities, then expanded and conquered surrounding territories, threats of invasion by enemy tribes or cities grew, too. Rulers surrounded their territories with walls and towers and moats and drawbridges. Cities themselves became shut in after dark.

Mesopotamian literature mentions locking doors and city gates after dark, and a bridge over the Euphrates River was equipped with removable beams to make it impassable at night.[12] In medieval European cities it wasn't unusual to hear a curfew bell (the word comes from the French for cover fire) at the end of the day commanding citizens to cease their trades and labours and return to their homes, lock their doors and shutter their windows. In some cities residents were even required to deposit their keys with the local authorities overnight.[13] Someone venturing out after curfew in medieval Copenhagen, Nuremberg or Parma would have to navigate massive iron chains stretched at waist height across those cities' streets, while in Moscow they would encounter logs rolled into the main thoroughfares. Curfews got later over time, from dusk, to eight, to nine and eventually ten o'clock, before being dropped altogether. That is, unless you were a woman, a beggar, a Jew or a foreigner, in which case you could still be arrested for being out and about at night.

Permission or not, a person abroad in a city at night could well be in danger. The satirist Juvenal spoke of the dangers of walking the streets of Rome after hours. One might get cracked on the head by a discarded cooking pot tossed from an upper window, he mused, or run down by a party of nobles hurrying in the other direction.

In London, theft was rife at night. Gangs of thieves were not averse to knocking their victims down with a hammer before lifting their purse or handkerchief. Torchbearers known as link-men were available to escort travellers around the city at night, but they could be as dangerous as the villains. Highwaymen would happily waylay a carriage and strip its occupants of their valuables. In seventeenth-century London night-time roads were so dangerous that, to travel the mile or so from Mayfair to Kensington, people would gather at a meeting point at the sound of a bell and walk in the safety of a group along the route known as Rotten Row.

Night-time fires were also a real threat, especially in towns and cities. The flames that brought light and comfort to the home could also wreak havoc in the overcrowded, higgledy-piggledy jumble of dwellings, stables, stores and workshops that characterised urban life. Once it took hold, a fire could easily spread from one building to another, timber to timber, destroying neighbourhoods or even entire cities. Catastrophic city fires in Rome (64 CE), Constantinople (406), Amsterdam (1421 and 1452), Oslo (1624) and London (1666) attest to their danger.

Nightwatchmen (not to be confused with nightmen, who emptied cesspits) were a common feature of pre-industrial urban life. Clay tablets found in the world's earliest cities in Mesopotamia contain poems referring to certain gods as 'night watchmen', who guided people around the streets at night.[14] Their roles ranged from policeman, in the days before police forces existed, to fire scout, sentry and timekeeper. Watchman duty was often a 'voluntary' civic duty and some cities locked the 'volunteers' into their post to prevent them slipping home to bed.[15] By the 1500s those who could afford it paid a substitute or a fine to avoid their turn, and cities were forced to put watchmen on the payroll.

Fear of night-time was not confined to rational dangers. In England, mischievous fairies danced around mushroom circles that could transport mortals into the netherworld if they happened to stray into them. Will-o'-the-wisps held aloft mysterious lanterns that misguided night-time travellers into deadly bogs and marshes. In parts of Eastern Europe, vampires emerged at night to suck the blood of innocents, while werewolves were known to roam as far west as Paris. Witches plied their spiteful trade all over Europe and in the new colonies of America. All these fiendish monsters were nocturnal, shunning the light (and in some cases destroyed by it).

Eventually, larger cities and some towns began to experiment with street lighting. For a while householders were obliged to keep a lit candle or lantern on their street-facing windowsill or hang a light on a pole from an upper window. By the late seventeenth century the first street lamps, fuelled by oil, appeared in Paris, Amsterdam, Berlin, London and Vienna.[20] When King William III moved his court to Kensington Palace in 1690 he had 300 oil lamps installed along the road between St James's Palace and Kensington, making the notorious Rotten Row the first artificially lit highway in England.[21]

Natural Lanterns

In the countryside, where there was less to fear in the way of vagabonds and thieves, leaving home after dark was not uncommon. People made use of natural advantages to navigate after nightfall, most importantly moonlight. Annually published almanacs allowed literate households to plan evening events based on tables of the Moon's monthly cycles.[16]

A full Moon always rises with the sunset and sets with the sunrise, making these nights the best time for travelling (a full

Moon was often known as the 'parish lantern', and parties and meetings were often scheduled for these dates). In Birmingham, a group of prominent figures of the late eighteenth century, including Erasmus Darwin (grandfather of Charles), Joseph Priestley, and early industrialists Matthew Boulton and James Watt, gathered once a month to discuss the issues of the day. Their meetings always coincided with the full Moon, giving rise to their name: the Lunar Society.

Natural features were also used to navigate after dark. In England's chalky South Downs people piled white chalk into large mounds known as 'down lanterns' to guide local travellers. In forests it was common to cut away tree bark to reveal the lighter trunk underneath, creating a luminous night-time corridor through the woods.

People also called on their other senses to guide them through the night. In Scottish lore 'the day has eyes, the night has ears', while in East Yorkshire idiom 'dark' is a verb meaning 'listen'.[17] In *A Midsummer Night's Dream* Hermia remarks that while dark 'doth impair the seeing sense, It pays the hearing double recompense'.[18]

To a careful listener, the crunch, pad, splash or clop of a foot-print could identify the surface being traversed. Features such as trees and water revealed themselves by sound, while rain falling or wind blowing transferred darkness into a three-dimensional space.[19] The screech and hoot of a pair of owls, the croak of a frog, the bleat of a sheep or the snort of a horse could all orien-tate a traveller, while following the sound of dogs barking was a good way to find the settlement they guarded.

Smells gave useful signals, too. Countrymen were familiar with the scents of various plants, and their geographical significance. The acrid smell of nettles crushed underfoot, amplified at night, indicated human settlement nearby, as did the welcome smell of wood smoke or cooking meat. Livestock and wild animals revealed their presence to the attuned nose, while dank and sulphurous odours warned of the dangers of bog and marsh.

Europe, 1700s

By 1700, while city nights were slowly becoming brighter, lights were going on elsewhere, too. This was the time of intellectual exploration and discovery known as the Enlightenment. The scientific and philosophical discoveries of Galileo, Bacon and Newton in the previous century had revealed there was much yet to learn about the workings of nature and the universe. One phenomenon that remained a mystery, despite its long association with humanity, was fire. As the ancient art of alchemy evolved into the modern science of chemistry, philosophers sought a more concrete explanation. This proved a considerable challenge.

For many years nascent chemists pursued the theory that fire comes from a substance called *phlogiston* (from an ancient Greek word meaning burning). Wood, they believed, comprised ash and phlogiston. When wood burned, it was *dephlogisticated*, leaving just the ash.

An important breakthrough occurred in 1774 when the British cleric Joseph Priestley discovered that air is not a substance in itself but a mixture of different gases.* He identified a gas he called 'dephlogisticated air' that supported combustion especially well and kept mice alive for longer than ordinary

---

* A Flemish chemist called Jan Baptiste van Helmont had identified that carbon dioxide was distinct from air in 1648, and coined the term gas (from the ancient Greek *khaos*), but it was more than a century before the enormous ramifications of his insight were understood.

air. The great French chemist Antoine Lavoisier called this gas 'oxygen' and explained its critical role in combustion, eventually overturning the long-held phlogiston theory.

Inspired by the new understanding of fire, in 1780 a Swiss chemist called Aime Argand came up with the first significant improvement in lighting technology for 40,000 years. Argand invented an oil lamp with a tubular wick that provided the flame much more air, and a glass chimney that kept the flame steady. His lamp was as bright as up to ten candles and burned oil far more efficiently. Argand brought his lamp to the English industrialist Matthew Boulton, co-owner of the Watt and Boulton steam engine manufactory in Birmingham and member of the Lunar Society. After various delays, the Argand lamp received a British patent in 1784, but meanwhile the market had become flooded with imitations and the patent was eventually quashed. Argand was devastated and eventually died in poverty, but his name lived on in the Argand lamp, the main form of domestic lighting for much of the nineteenth century.[22]

A decade later a Scotsman called William Murdoch invented a new form of lighting that did away with wick altogether. People had known for years that certain types of coal gave off a gas when heated, and that this gas gave off a bright white light if ignited. After many experiments, Murdoch turned this phenomenon to useful effect. He set up a small plant behind his house to heat the coal, and rigged up a system of pipes from the plant to the house, and a series of pipes and apertures on the walls. The plant produced coal gas, and this was piped inside. When lit, the gas flowing through the apertures provided a warm white light with not a wick or drop of oil to be seen.[23]

As it happened, Murdoch was an employee of Watt and Boulton, and he continued his experiments with gaslight at their factory in Birmingham. In 1802 he lit up the factory's exterior in a splendid display 'that showed off the brilliance and versatility of gaslight, causing much excitement amongst the local folk'.[24]

Murdoch followed this by installing a gas lighting system at one of the largest cotton mills in the country, Philips and Lee's in Manchester, in 1805.[25] For the first time, a large space could be lit, brightly and (relatively) safely, and without the frequent tending of candles or lamps. Over the next decades, many other mills and manufacturing workshops followed.

Watt and Boulton, perhaps bruised by the Argand lamp fiasco, ignored Murdoch's pleas to apply for a patent for his gaslight invention, and this allowed other enterprising souls to get into the game. A German called Frederick Winsor realised gaslight's potential to be supplied on a large scale from centralised plants. Focusing his efforts on London, he built a display of gaslight in Pall Mall, and in 1812 established the world's first gas lighting company. Gasworks and gasometers sprung up in strategic London locations, and mains were laid connecting the gas supply to streets, homes and shops in the vicinity. By 1823 more than 60,000 private (home) lights and 7,000 street lights were being supplied by the four largest companies in London.[26] By the late 1820s almost every medium-sized town was supplied with gaslight.[27]

Northern England, 1800s

Before the invention of gaslight most things were made by tradesmen in workshops or by putting out 'piecework' to home-based artisans. Gaslight facilitated the transformation of manufacturing into the factory system. While candles and lamps were sufficient for small-scale works, they were impractical and expensive for lighting larger operations. With gaslight, the scale of manufacturing could be multiplied several times, with consequent improvements in efficiency and profit. Large-scale machinery could be operated by workforces organised into specialised tasks. With reliable lighting available, the machines could be run to fixed schedules all year round.

Small-scale workshops were replaced by vast factories operating production lines, and tradesmen and artisans were replaced by minimally skilled workers who became interchangeable units of production, just as the components they made were becoming interchangeable parts.

With gaslight, the dark satanic mills William Blake described in his 1804 poem *Jerusalem* were eventually made dark no more. Whether they were less satanic is debatable. Children still made up a significant part of the workforce. Working conditions were harsh and punishments for transgressions were severe. Shifts were long and wages were low. But for all this, farming families displaced by changes in agricultural methods flocked to the growing cities and the prospect of factory work.

Gaslight brought the night abruptly into the productive sphere. It eroded the primal dichotomy between night-time and daytime that had existed since people started gathering around campfires. Industrialisation required efficiency – the hallmark of daytime activities – not firelight's historical ally, imagination.

Homes embraced gaslight, too. Wall sconces and chandeliers were replaced with gas fittings, and newly built homes were plumbed for gaslight – the elaborate ceiling roses one sees in Victorian-era properties disguised gas ventilation grilles. Gaslight made reading at night much more comfortable, and, as books and magazines were more affordable thanks to improved printing methods, publishing boomed. Literacy rates soared throughout the century, from fewer than 40 per cent of women and around 60 per cent of men being able to read in England in 1800, to nearly 70 per cent of women and 80 per cent of men by 1860, and near universal literacy by 1900.[28] Families no longer sat listening to one member reading aloud by the light of a lone candle. With gaslight, there was sufficient light for everyone to read silently to themselves.

As well as being brighter, gaslight was instant, adjustable and apparently inexhaustible. It removed the link between fuel and flame, instigating an interesting psychological shift. With the gas supply coming from a centralised source, households were no longer independent, self-sufficient units, making and providing their own light. They were tied to a remote provider, dependent on an external source. If the gas supply went down, as it often did, the whole neighbourhood's lighting went down. This was the beginning of life on the grid.

By the mid-1800s lighting had been industrialised. It was now a large-scale operation mass-producing a uniform product. Households connected to gas were also industrialised. As individuals became units of production in factories, their homes became units of consumption.

Once gaslight appeared, it became both safer and socially acceptable to travel the streets at night. The streets themselves became the entertainment, as couples strolled about inspecting the beautifully lit merchandise in the newly glazed and gaslit shops of the day. Music hall and melodramas became popular, and theatres sprung up in London's West End and New York's Broadway.

It wasn't just frivolous activities that flourished under gaslight. Once city streets were safe, ordinary people went out in the evenings to attend public talks, lectures and meetings about matters of interest or concern, accompanied by cheap and accessible pamphlets and flyers. This made the early to mid-1800s a time of great political and social momentum: 400 years after King Henry VI established that only male property owners could vote, a series of Reform Acts starting in 1832 gradually extended suffrage to more and more men (and, much later, women). Parallel movements around the UK established schools for poor children, leading to state funding of schools by the mid-century. Industrial reforms were introduced, trades unions were legalised, and the Factory Acts of the 1830s and '40s restricted working hours and legislated for improved conditions for factory workers.

The structure of society shifted from a series of independent entities operating locally, to fewer, but much broader and interdependent, networks operating across neighbourhoods, cities, and even nations. The same broad trend could be seen in life at work, at home, and in society at large. Seeing in the dark with gaslight was making the world both larger, opening up new opportunities for many people, and smaller, as each person became a bit player in a much larger scene, rather than the star of their own small drama.

The Early Electricians

As gaslight was transforming the spaces and lifestyles of the nineteenth century, a quest was under way to understand an invisible force that had puzzled natural philosophers for centuries. It took decades to reveal the characteristics of this enigmatic effect, and decades more to harness it. Once subjugated, however, it transformed not only the way the world was lit, but almost every aspect of human life.

Around 600 BCE the Greek philosopher Thales of Miletus wrote of the propensity of a piece of amber rubbed with fur to emit sparks and attract small objects. More than 2,000 years later a prominent physician and philosopher called William Gilbert conducted a series of 'attraction' experiments and dubbed materials that attracted when friction was applied 'electric', from the Greek word for amber, *electron*.[29]

Various other European 'electricians' experimented with the attractive and repulsive powers of 'electrics' over the next few decades, including Sir Francis Bacon, Robert Boyle and Sir Isaac Newton. Little progress was made towards understanding electricity, but practitioners devised ever more elaborate demonstrations of its properties. Feathers and leaves floated miraculously towards or away from charged items; charged fingers turned pages without touching; men's hair was made to stand on end and ladies' cheeks to snap with an electric shock when kissed.

Benjamin Franklin became interested in electricity in the 1740s.[30] In June 1752 he performed his famous experiment, flying a kite in a thunderstorm. He 'presented his knuckle to the key [tied to a wet kite string] and … perceived a very evident electric spark'.[31] He managed to complete his lightning experiments without killing himself (others were not so lucky), but in the end his famous discovery didn't take the theoretical understanding or practical application of electricity much further forward.

Half a century later, around 1800, Italian Alessandro Volta created the chemical battery, for the first time producing electricity in the form of a current rather than a flash. British chemist Sir Humphrey Davy took Volta's idea forward and built an enormous battery in the basement of the Royal Institution in London. In one of his popular evening lectures he gave the first public demonstration of electric current – a fizzing and crackling arc of brilliant white light produced by the current leaping between two charcoal sticks.

Davy called the effect the arc lamp. He had created the first man-made electric light, but further progress was hampered by the inability of chemical batteries to generate sufficient electricity to sustain a current. Even the largest batteries could only provide a short burst of current to feed the power-thirsty arc flame.

Michael Faraday, Davy's successor at the Royal Institution, changed everything in 1831 when he discovered that a magnet moving through a coil of wire, or a coil moving within a magnetic field, produced an electric current. This provided a means to produce an electric current using mechanics rather than chemistry: mechanical steam engines could power continuous movement, and this could produce a continuing flow of current, resulting at last in a practical electric generator.

Faraday's experiments into magnetism, electricity and electromagnetic forces changed the way physicists saw the world. Newton's General Theory of Gravity had assumed particles and objects attract one another in proportion to their size and distance across an empty space. Faraday proposed that the space between objects was not empty, but instead filled with invisible waves and forces. He was convinced these waves and forces were somehow connected to light, too, and eventually proved it. His later experiments demonstrated that the speed of electric transmission was not instantaneous, as had been believed, but occurred at the speed of light. Faraday's insights were critical building blocks in Einstein's General Theory of Relativity.

Exactly 400 years after Jan Van Eyck's Arnolfini portrait signalled a shift in worldview from belief and imagination to evidence and observation, the world was entering a new Invisible Age, governed by science.

Europe, Late 1800s

It took another seventy years, but eventually a usable version of Davy's arc light was unveiled in 1878. During the Paris Exposition Universelle that year, sixty-four electric arc lights were installed around L'Opéra, powered by three steam-driven generators. In London in the same year, the Victoria Embankment was lit with arc lights, and two years later America's first arc lights were switched on in Wabash, Indiana.[32]

Arc lights were so bright they had to be installed well above the line of vision, much higher than gaslights had been. Some US cities did away with streetlamps and installed great central towers that lit entire neighbourhoods and even cities. Some 122 towers, each around 50m high, lit Detroit. A French visitor to San Jose, California, commented that electric light from several 60m-high light towers 'inundates the whole town'.[33]

The brightness and uniformity of arc lights were seen variously as liberating and tyrannical. On the one hand they were believed to enhance law and order, as the dark alleyways where nefarious deeds were done became as exposed to scrutiny as the high street. Some felt they were democratic, as their glare reached poorer neighbourhoods not served by the gas companies.[34] Others, however, found the uniform arc light oppressive. H.G. Wells described 'gigantic globes of cool white light [that] shamed the pale sunbeams'[35] and Robert Louis Stevenson made an eloquent plea for the return of gas streetlamps on aesthetic grounds after seeing the electric arc lights in London and Paris, complaining that:

> A new sort of urban star now shines out nightly, horrible, unearthly, obnoxious to the human eye; a lamp for a nightmare! … Mankind, you would have thought, might have remained content with what Prometheus stole for them and not gone fishing the profound heaven with kites to catch and domesticate the wildfire of the storm.[36]

Despite Stevenson's comments, electric light was far from domesticated at this stage. Arc lighting was simply too bright. It needed to be remade in a smaller, calmer form, just as the cosy wick remade the brutal torch thousands of years before.

The solution was a different form of electric light altogether. Incandescence is the light that comes from a white-hot material. As the commercial potential of arc lighting was being developed, electricians in Europe and America were searching for a practical way to turn electricity into incandescent light.

Unlike most previous scientific races, when the finishing post was the honour of priority among scientific peers, the latter part of the battle to master electricity was fought at the British and US patent offices. As the developed world continued to industrialise throughout the nineteenth century, the importance of intellectual property – an idea as invisible as electricity but just as important to the modern world – had continued to grow. In fields where entrepreneurs sniffed potential profit, the quest for scientific discovery often gave way to the quest for patentable invention: new products or techniques that could be owned, protected and exploited commercially. Despite its reluctance to be subjugated, many could see that electricity had enormous commercial potential. Gas lighting was a huge success, but its light was hot, prone to occasional accidental fire, and took a lot of oxygen from the rooms it lit. If cool, clean electric light could be provided economically, the potential rewards were colossal.

By the mid-1800s describing the underlying theory of electricity had become less important than coming up with practical applications for it. Experimenters of this period were not necessarily physicists, nor even scientifically trained, but all manner of ingenious self-starters. The objective was clear – an economical, incandescent electric lighting system – and the stakes could hardly be higher.

The problem with incandescence was the high likelihood that the heated filament would burst into flames. The obvious solution was to situate it within a vacuum where combustion was impossible – sealed in a glass bulb, perhaps. This was easier said than done, however. Patents for incandescent bulbs were filed (by different people) in 1841, 1845 and 1856, but none ultimately proved viable, mainly due to the difficulty of achieving a true vacuum.[37] This problem wasn't solved until an efficient vacuum pump was developed in 1865.

The next hurdle was finding a suitable material for the filament. Experimenters tried dozens of different materials over several decades. English self-taught chemist and Newcastle businessman Joseph Swan experimented for years with carbonised paper and cardboard filaments until finally, in late 1878, he demonstrated his successful light bulb to the public. Swan patented several of his lamp's components, but did not patent the lamp itself, believing it to be an item that was already well known.[38] This was to prove a major miscalculation.

New Jersey, USA, 1878

On the other side of the Atlantic, much interest and activity was directed towards electricity. Several companies were working on perfecting arc lights for public spaces, and the popular press was abuzz with the potential of this mysterious new energy source. Several inventors were trying to develop an incandescent lamp, struggling with the same practical problems as their European counterparts. Gas lighting was big business by this stage, and the owners of the gas companies were some of the richest men of their time: Cornelius Vanderbilt, J. Pierpont Morgan, and William Rockefeller. They were keen to protect their gas interests, but equally keen to not miss the next big thing.

In the autumn of 1878, shortly after the display of arc lighting at the Paris Exposition, and just as Joseph Swan was preparing to display his electric light bulb in England, another name joined the race. On 8 September the *New York Herald* printed a claim by Thomas Edison that he would have a workable electric lamp within six weeks, and half a million lamps installed in New York City homes within a year.[39]

Thomas Alva Edison was a lifelong experimenter. A laboratory in his bedroom was moved to the basement of his childhood home after various mishaps. He was fired from his first job – selling newspapers and sweets on the local railway line – when the laboratory he had set up in the train's baggage compartment exploded.

By 1878 he was no longer tinkering around in basements and backrooms. Edison was the pioneer of industrialised creativity, called the inventor of invention. He had built the world's first commercial research and development lab in Menlo Park, New Jersey, and filled it with 'as interesting an aggregation of learned men, cranks, enthusiasts, plain muckers as ever gathered under one roof'.[40] His quadruplex telegraph, patented in 1874, had transformed global communications by allowing large volumes of news and information to be sent instantaneously.* On Christmas Eve 1877, he had filed a patent application for the phonograph, the first device ever to both record and reproduce sound. In 1878 Edison was being called the Wizard of Menlo Park and was a household name on both sides of the Atlantic.

As well as being an inventor, Edison was an ambitious entrepreneur. He was backed by J.P. Morgan and connected to many of the other mighty capitalists of the period. Edison thought big. He was interested in the potential of electricity as a whole system, from its generation to end uses, unlike most of his competitors whose attention was focused on the electric lamp. He was already thinking beyond the lamp to potential electric appliances for household chores such as sewing, cooking and heating.[41]

---

* Samuel Morse invented the telegraph in the 1830s.

When Edison made his boast to the *Herald* in 1878 he hadn't even started on the incandescence problem, much less the broader question of creating a viable electricity generation and distribution system. Nevertheless, such was the perceived power of Edison and his team that the price of gas stocks tumbled on hearing the news.[42] They soon recovered, however, when no Edison lamp materialised.

Edison eventually gave a public display of incandescent lamps a year later, in December 1879. While many histories mark this event as the moment the light bulb was invented, in reality Edison and his team drew heavily on the failures of others, either in their failed experiments or their inadequate patent protection. Challenges to Edison's priority went on for years, and many of his initial claims were eventually refuted. In the end, though, Edison's canny business sense and powerful backers ensured he had the final victory.

Over the next decade, among the rough and tumble of late nineteenth-century New York entrepreneurialism, Edison made good on his boast and developed a full-blown electrical delivery system. He wired the mansions and yachts of his wealthy friends, and switched on his first neighbourhood lighting system in the summer of 1882, near Wall Street in New York City. A local reporter noted at the time: 'the dim flicker of gas … was supplanted by a steady glare, bright and mellow, which illuminated interiors and shone through windows fixed and unwavering.'[43]

Meanwhile in Great Britain, Joseph Swan founded the Swan Electric Light Company in 1881 and installed around 1,200 incandescent bulbs in Richard D'Oyly Carte's state-of-the-art Savoy Theatre in London's West End, making it the world's first public building lit entirely by electricity. Edison promptly sued him for patent infringement, and after some legal wrangling the two companies merged.

The era of electric lighting had begun, and the Electronic Age was just around the corner.

The Electronic Age

Chicago's Columbian Exposition of 1893 used 200,000 incandescent bulbs and 6,000 arc lights to illuminate the purpose-built White City. As yet, its displays weren't real life, but they held the promise of an illuminated future, a world of gentle brightness at the flick of the switch. This was a luxury not even dreamed of in the past. The visiting public didn't know it yet, but these dramatic displays of light were merely the prelude to a whole raft of electronic consumer products that would utterly transform their lives. Within a couple of years a series of electrical appliances was added to the electric lamp, beginning with the iron, the fan and the kettle.

Electricity changed the way people saw themselves, their lives and the world. Gas lighting had removed the direct link between people and firelight and separated the flame providing the light from the fuel that powered it. Electricity was the final step away from firelight and onto the grid: the flame was gone altogether and only the light remained, served by an invisible current wired into the home from an invisible power station somewhere beyond. The householder soon didn't even see the light – didn't think about it beyond flicking the switch. They trusted the electric company to provide a continual flow of current, and science to understand the mysterious workings of electricity on their behalf.

The capitalists who financed electricity's development saw the world on a different scale from most people. They saw the potential of a vast infrastructure serving thousands or even millions of individual units to make profits, and they had the means to achieve it. The risks were huge but so was the potential reward, as had been proved already by the railroads, the telegraph and gas lighting. These titans saw from the outset that electricity was ripe for the same treatment if they could win the technology race.

Edison's domination of electrification in the US and Europe shifted the world on its axis. Until the second half of the nineteenth century, Europe was the epicentre of science, and Great Britain was the centre of invention and the world's industrial powerhouse. The pre-eminence of the USA in the development of electric light and electricity propelled it to the status of economic superpower. As the Electronic Age progressed, the precocious New World eclipsed Old World Europe and came to dominate the world.

Light on Tap

When God said 'Let there be light', He only went halfway. Modern industry has finished the job.

Life on Earth has never been brighter. Today, there are around 12 billion light bulbs in the world and more than a billion street lights. Compared with our predecessors, we are oblivious to night, to the cycles of the Moon and to the seasons. We hardly use the 240 million rods in our eyes because we are constantly surrounded by artificial light, and we spend our waking hours feasting on illuminated electronic screens. Many people never experience true darkness or see the Milky Way. Indeed, we now talk about light 'pollution', and passionate campaigners around the world fight to protect the precious few remaining 'dark skies'.[44]

For most of us, however, it is only on the very rare occasions that we find ourselves without artificial light that we even notice that it's there.

SHOWING

MASS MEDIA AND THE CONQUEST OF SEEING

12

NATURE'S PENCIL:
PHOTOGRAPHY

They differ in all respects, and as widely as possible, in their origin,
from plates of the ordinary kind, which owe their existence to the
united skill of the Artist and the Engraver.
They are impressed by Nature's hand.

Henry Fox Talbot (1800–77), *The Pencil of Nature*, 1844

Like London buses, similar inventions often appear at the same
time. We can put this down to zeitgeist, a nineteenth-century
German concept meaning the 'spirit of the age' that seems to
make some inventions almost inevitable. As the pioneering historian of psychology Edwin Boring put it in 1950:

> not only is a new discovery seldom made until the times are
> ready for it, but again and again it turns out to have been
> anticipated, inadequately perhaps but nevertheless explicitly,
> as the times were beginning to be ready for it.[1]

By the early 1800s, the Western world was beginning to be ready
for photography. Developments in science, philosophy, art, society and industry combined to foster a spirit of the age in which

several actors, unknown to one another, started looking for ways to combine optics and chemistry and 'fix' images automatically.

Europe, Early 1800s

For two centuries after the invention of the telescope, interest in observing and chronicling the workings of nature continued to rise. Scholarly endeavours placed observation and direct experience at the forefront of knowledge and ideas. Sir Isaac Newton's insights in the late 1600s showed that the natural world was not guided by invisible, animate and sacred spirits, but was rather a giant, quasi-mechanical system governed by a series of rules and laws that could only be understood – and indeed were slowly being revealed – by keen observation and methodical experimentation. Philosophers such as Descartes and Locke had overturned the classical view that certain innate truths exist. Knowledge was not innate, the newly modern scientists believed, but must be acquired by direct experience and sensory perception. I see, therefore I know.

The eighteenth and nineteenth centuries were times of enquiry and exploration. Scrutiny, surveillance, investigation, classification and analysis of nature were taking place all over the world – from the mapping of the stars made possible with enormous new telescopes, to ambitious geographical and geological surveys, to the taxonomy of plants, animals and insects, to studies of the invisible world with microscopes. The appetite to see and record the physical world was voracious.

Intrinsic to the need to experience and observe was travel, and this took place on a grand scale during this period. The great new colonial powers sent fleets of ships around the world's oceans in pursuit of trade, and overland expeditions into the interiors of the Indian sub-continent and Africa.

Newly independent Americans were exploring and surveying the continent, pushing ever further west against considerable – but ultimately unsuccessful – indigenous resistance. In Europe, wealthy and aristocratic young men – and, sometimes, women – routinely embarked on the Grand Tour, a lavish precursor to the modern gap year. These fortunate souls and their retinues spent months exploring the great cities and sites of Europe, observing and absorbing the great works of antiquity and Western culture, and hobnobbing with their international contemporaries. Seeing the glories and treasures of the past with their own eyes would, they believed, imbue them with the civilising powers of their European antecedents.

All this observing drove a requirement to record what was seen. The options available were sketching, drawing and painting, and topographical, scientific and botanical illustration were popular career paths for artists at this time. Joseph Banks employed the Scottish artist Sydney Parkinson to join him on Captain James Cook's voyage to the Pacific in 1768, and his drawings are still celebrated as the first glimpse of the exotic flora and fauna of the region. Maria Sibylle Merian brought the first glimpses of the New World back to Europe in her drawings of Surinam wildlife, while the French illustrator Claude Aubriet did the same thing for the Middle East.[2]

Rigorous depiction was all very well if one were a talented draftsman, but not all travellers were so gifted. Scholars and amateur enthusiasts – as well as artists – turned to the *camera obscura* to help them capture accurate representations of landscapes and objects. Just to remind you, the *camera obscura* is a natural optical phenomenon whereby the image of a brightly lit scene can be projected through a small hole onto a surface in a darkened space. The original *camera obscuras* were actual rooms, but by the eighteenth century there were portable versions available. The so-called Royal Delineator was a mahogany box with a telescopic viewfinder fitted with two convex lenses, a tilted mirror to redirect the image to the top of the box, and a lens at the top to enlarge the image on the drawing surface.[3] It received a king's patent around 1780 and was reputedly favoured by famous painters including Canaletto and Sir Joshua Reynolds as well as wealthy amateurs. Another contraption known as the 'field camera' perched on top of a small, dark tent in which the artist sat tracing the image projected and reflected down onto a piece of paper.[4]

In 1806 an Englishman called William Hyde Wollaston, despairing at his lack of ability to capture the grandeur of the Lake District on paper, patented a new optical drawing aid he called the *camera lucida*. This was an elegantly simple and highly portable device that used a small prism suspended on a stand to reflect the image being observed in such a way that it appeared to float over the paper like a ghost, allowing the user to trace the subject in as much detail as they chose. It quickly became the preferred instrument to aid observational drawing and was used by artists and amateurs all around the world for the first half of the nineteenth century.[5]

Around the same time as Wollaston launched his device, the art world was experiencing a quiet revolution. With some notable exceptions, landscape painting had until now been seen as the poor relation of art, the precinct of jobbing topographers recording monuments for the tourist trade and country seats for the vanity of provincial squires. A Suffolk countryman called John Constable (1776–1837) and his Londoner contemporary J.M.W. Turner (1755–1851) changed that view and, in the process, paved the way for more radical artistic changes in the second half of the nineteenth century. Both artists approached nature in all its forms – embracing such mundane details as 'willows, old rotten banks, slimy posts' as well as the more dramatic atmospheric effects of light and weather – as a subject worthy of their full attention. This was a departure from traditional landscapes, where it was typically man's interventions in nature that took centre stage. Constable and Turner used pioneering techniques of colour, brushwork and composition to represent nature as they saw it instead of as the artistic academies deemed it. They painted with the eye rather than the mind.[6]

There were also major advances in the new field of chemistry in the eighteenth century. In 1727 a German chemist observed that compounds of silver turned dark when exposed to light.[7] Later that century, the great chemical rivals Lavoisier and Priestley established that chemicals reacted in predictable, consistent ways when combined.

In 1796 another German used chemical reactions to devise a printing technique he called lithography, meaning drawing with stone. Lithography could reproduce images faster and more cheaply than earlier etching or woodblock printing, and made graphic art and pictures widely available to ordinary people for the first time.[8] A new market for printed pictures reproduced from paintings or made especially for the purpose of printing developed to serve the middle classes.[9] Reproduced images became commonplace for the first time.

In manufacturing, the Industrial Revolution was well under way, and more and more traditional tasks and trades were being industrialised and automated by the new machines. Could the artist's craft go the same way?

The currents of scientific, artistic, industrial and social change were mixing and converging. The zeitgeist was slowly forming an idea that hung in the atmosphere like the fragmentary memory of a dream: could an image be taken directly from life and captured, frozen and fixed forever?

England, 1800

Thomas Wedgwood, son of the famous potter Josiah, was the first person to try to capture permanent images by combining the optical effects of the *camera obscura* with chemical reactions to light. Thomas was a sickly young man who seems to have suffered from a crippling form of depression, but as a gentleman of means and standing he was acquainted with many leading figures of the day who were, or were on the way to becoming, prominent in art and science.[10]

As early as 1800, Wedgwood experimented with making what he called 'automatic' images. He held a glass plate with a picture painted on it over white paper impregnated with silver nitrate solution and exposed it to the Sun. He found that 'the rays transmitted through the differently painted surfaces produce distinct tints of brown or black, sensibly differing in intensity according to the shades of the picture'.[11] The problem was that as soon as his 'automatic' images were exposed to more light they went black. Wedgwood also tried but didn't manage to make a visible image with a *camera obscura*.

Wedgwood's friend, Humphrey Davy – the same Davy who later demonstrated the first electric light – wrote up Thomas's experiments with light and images in a paper entitled 'An Account of a Method of Copying Paintings upon Glass, and of Making Profiles, by the Agency of Light upon Nitrate of Silver, with observations by Humphrey Davy. Invented by T. Wedgwood, Esq'. It appeared in the very first journal of the new scientific organisation, the Royal Institution, in 1802. It seems neither Wedgwood nor Davy fully appreciated the potential significance of their discovery. Wedgwood died in 1805, leaving no reference to his photographic experiments in his letters or papers.[12] Davy became a famous chemist but he never returned to the problem of 'fixing' photographic images. Wedgwood's significant step towards the invention of photography was all but forgotten for decades, leaving others to start from scratch to find ways to draw with light.[13]

Le Gras, France, 1826

Like Thomas Wedgwood, young Joseph Nicéphore Niépce came from a well-to-do family with a keen interest in science and technology. In 1813, Nicéphore Niépce turned his attention to the newly popular printing process, lithography. Niépce couldn't draw, so he began to experiment with ways of producing images automatically. He trained a *camera obscura* on the view from his window and tried using various surfaces treated with different chemical compounds to reproduce the projected image. In 1816 he achieved a negative image on paper treated with silver salts – the first image ever produced by a camera – but like Wedgwood's earlier attempts it turned black when exposed to more light. Niépce shifted his efforts to attempting to obtain a positive image.

In 1826 Niépce finally succeeded in capturing the world's first permanent photographic image. He called it a heliotype, or 'sun drawing'. It was the scene from his window, projected through a *camera obscura* over several sunny days onto a pewter plate treated with bitumen of Judea.

Soon after, Niépce met an artist and theatre designer called Louis Daguerre. Daguerre was a master showman with grand ambitions, and was already famous for having created a spectacle called the Diorama. This was a purpose-built theatre that used multiple sets and lighting effects to create dramatic scenes from nature and the ancient world, and it had been a sensational success in Paris and London. Daguerre was fascinated by the idea of capturing images directly from life and was keen to get involved with Niépce. In December 1829, Niépce and Daguerre entered into a ten-year partnership to develop heliography commercially. When Niépce died suddenly in 1833, Daguerre continued his work, and eventually came up with a process that created a permanent, extremely sharp image on a silver-coated sheet of copper. Daguerre was now ready to seek investors, and he set off to Paris to demonstrate his invention, humbly named the daguerreotype, to leading figures in the arts and sciences.

When commercial investors weren't forthcoming, Daguerre found a charismatic champion in the astronomer and politician François Arago. On 7 January 1839, at the French Académie des Sciences, Arago announced:

> M Daguerre has devised special plates on which the optical image leaves a perfect imprint – plates on which the image is reproduced down to the most minute details with unbelievable exactitude and finesse.[14]

Arago proposed that, rather than issue a patent, the French Government should buy the invention from Daguerre and give it freely to the rest of the world in the name of progress and to the glory of France.[15]

According to contemporary reports, 'the entire civilised world was stunned by the announcement and Frenchmen went wild'.[16] Not least among these was Nicéphore Niépce's son, Isodore, who was incensed. He believed his father's contribution to the invention had been marginalised, and that his own inherited share in the partnership was also being sidelined. He successfully sued Daguerre, delaying the release of any further information about the daguerreotype until the court action was resolved.

England, 1839

Another individual stunned by Arago's announcement was an Englishman called William Henry Fox Talbot. Talbot was yet another provincial gentleman of means, a polymath, inventor and scientific tinkerer, but no artist. Whilst on honeymoon on Italy's Lake Como in October 1833 he tried to make some sketches with a Wollaston *camera lucida*. However, 'when the eye was removed from the prism – in which all looked beautiful – I found that the faithless pencil had only left traces on the paper melancholy to behold'.[17]

A decade earlier he had had a similar lack of success drawing with a *camera obscura*, despite the beauty of its 'fairy pictures, creations of a moment, and destined as rapidly to fade away … leading him to wonder … if it were possible to cause these natural images to imprint themselves durably and remain fixed upon the paper!'[18]

In early 1834 Talbot began experimenting with what he called 'photogenic drawing'. He tried brushing solutions of silver compounds onto paper and exposing them to light. He produced some negative images – that is, the light areas were dark and the dark areas light – and after a while realised that if he could make the first image with transparent paper, he could use it to make a second image that would be reversed, and thus positive.[19] While this process required two stages, Talbot recognised the advantage of being able to use a single negative to make an unlimited number of positive images, and he pursued this line of experiment for the next couple of years. In early 1839 he was still grappling with the problem of how to stop his images fading away when he heard of Arago's announcement in Paris.

Talbot wasted no time. He was determined to publish his discoveries before Daguerre to avoid any claim he was influenced by the Frenchman's invention. He contacted the Royal Society immediately, and offered to present a paper on the matter entitled 'A New Art of Design', and he sent them examples of his photogenic drawings. Talbot's paper was read to the Society on 31 January, just three weeks after Arago's speech, and gave full details of his photographic process as it then stood.

Talbot acknowledged that his method was not yet commercially viable. As a gentleman, he was probably more interested in securing scientific priority than wealth, and said he was sharing his discoveries in order that others could improve on them, while he continued trying to improve them himself. He was convinced his process was both different from and superior to Daguerre's.

Talbot turned out to be right on both counts, but Wedgwood, Niépce, Daguerre and Talbot can each rightly claim a pivotal role in the discovery of photography. Wedgwood was the first to capture an image on a medium using light and chemicals; Niépce created the first fixed image; Daguerre produced the first commercially viable photographic process; and the negative–positive method Talbot perfected over the next couple of years eventually superseded the daguerreotype and became the photographic standard for a century and a half until the advent of digital cameras.

Nevertheless, in the spring and summer of 1839 it was obvious that Daguerre's process had a clear lead over Talbot's.

Paris, August 1839

On the afternoon of 19 August 1839, more than 200 people waited in the courtyard in front of the imposing pillared entrance to the Institut de France, on the left bank of the Seine facing the Pont des Arts and the Louvre. The crowd was the overflow that had accumulated since the last seat inside the building had been taken hours earlier. Inside, members of the Academies of Beaux-Arts and of Sciences took their places in the centre of the main chamber, surrounded by the packed visitors' stalls and overlooked by busts of France's most distinguished contributors to art, science and philosophy. At the focal point of the room sat Messrs Daguerre, Niépce (junior) and Arago.

At three o'clock M. Arago rose and addressed the crowd. He described the history of the *camera obscura* and discoveries of the light sensitivity of various substances, the early experiments of Thomas Wedgwood, the achievements of Nicéphore Niépce in fixing the first permanent image from nature, and the further trials Daguerre had carried out after Niépce's death.

Finally, M. Arago came to the part everyone was waiting for: the description of the daguerreotype process. In the event he gave a fairly abbreviated resumé, along with a promise to publish a full manual and present practical demonstrations soon. If the crowd were disappointed by the lack of detail, they didn't show it. Even the stern Academicians cheered, and the announcement was widely reported in Paris the next day and London three days later.[20]

Things moved fast from there. The daguerreotype process was made freely available to all thanks to the largesse of the French Government, but Daguerre had cannily retained a patent on the equipment required to execute it. By early September there were advertisements in the French press for the official daguerreotype manual, camera and plates. The manual was published in French on 7 September and within a week an English version was available in London. The news crossed the Atlantic as fast as the newest steamer ship, the *British Queen*, could carry it.[21] The French equipment licensee sent its agent, François Gouraud, to New York to stimulate interest and secure equipment orders. Demonstrations and exhibitions started appearing in big American cities within weeks, and newspapers and journals were full of discussion of the 'new art'. One editor gushed:

> All nature shall paint herself ... fields, rivers, trees, houses, plains, mountains, cities, shall all paint themselves at a bidding, and at a few moments' notice ... by virtue of the sun's patent, all nature, animate and inanimate, shall be henceforth its own painter, engraver, printer, and publisher.[22]

Nature's Portraitist

Among all the rapture and excitement around the launch of the daguerreotype there was almost no discussion of its potential to capture the human face. Daguerre's demonstration pictures included street scenes, still lifes, landscapes, exotic locations, details of sculptures and other works of art, botanical and biological subjects, even a picture of the Moon, but no portraits. Remarks made by Daguerre and other commentators at the time enthused at length about photography's potential as a tool to further the arts and science, but its potential social contribution was quite overlooked. This may have been because early daguerreotypes took a long time to expose, during which a human subject would be likely to move. In fact, early Daguerre streetscapes did include people but because they were usually in motion they didn't register on the plate. People, horses and carts moving through street scenes were invisible on the daguerreotype. The only humans to appear in any of Daguerre's images appeared accidentally: two tiny figures in a Paris street scene, a bootblack and his customer, captured by chance as two still figures in a dynamic – and therefore invisible – milieu.

In overlooking portraits, Daguerre had grossly misjudged the market. Others made no such mistake, and portraits started appearing almost immediately. In October or November 1839, very shortly after the first reports of the daguerreotype process reached Philadelphia, local chemist Robert Cornelius made himself a camera.[23] In the bright sun outside his family's shop he stood in front of his camera, uncovered the aperture, stood deadly still for a while, then carefully recovered the lens. The resulting image was the first known photographic portrait in history and also the first selfie. It captures a handsome young man with dishevelled hair looking slightly to the left of the camera with a suspicious expression. The picture portrays life and likeness like no painted or drawn portrait ever had, or ever could have, because it captured Cornelius exactly as he was for those fleeting few moments.

By the middle of the following year, two New Yorkers had opened the world's first photographic portrait studio on Broadway and the year after that Europe's first studio opened in Regent Street, London. The requirement to remain deadly still for several uncomfortable minutes was no impediment to demand, and studios provided modified chairs and strategically placed pedestals to help people hold their poses. Portraits were the killer app for the daguerreotype. Yes, people were impressed with photographs of far-off lands, and their town hall, and statues carved centuries earlier. But what they really wanted to see, and to own, was a picture of themselves. The daguerreotype obliged.

Men of all manner of backgrounds – chemists, tradespeople, artists, jewellers, merchants – began investing in their own daguerreotype equipment and setting up shop. In the US, an official Daguerre-endorsed camera – a wooden cube about the size of a filing box – cost $51, and each plate was $2, meaning an individual could set up in business for under $100. Selling pictures at $7 each, an entrepreneur could make back his invest-ment and get into profit within a few weeks.

One of the earliest and most successful photography entrepreneurs was the New Yorker Mathew Brady. He was one of the first people to appreciate the potential mass appeal of the celebrity photographic portrait, and set up a portrait studio on Broadway in 1844. Brady sought out the notable figures of the day and offered to photograph them for free, then displayed their portraits in a gallery for which he charged an entry fee. In 1850 he published a best-selling album of lithographs (taken from daguerreotypes) he called *The Gallery of Illustrious Americans*. Brady insisted that every image produced by his studio was marked with his name, and soon became the go-to photographer for celebrities and those aspiring so to be.

In February 1860 an unknown young political candidate from Illinois stopped at Brady's studio on his way to deliver a speech nearby. The candidate was tall and gangly with a long scrawny neck and rugged features. Brady pulled up the young man's collar to hide his neck and posed him standing just offset, with one hand placed lightly on a book as if taking an oath, and staring directly at the camera. The resulting image was a figure to be reckoned with: upstanding, elegant, trustworthy. It was printed in engraved form on the cover of the candidate's printed speech and, when he was elected nominee for president, it appeared in newspapers and on campaign buttons and flyers. When Abraham Lincoln, Brady's subject that day, was elected as the sixteenth President of the United States of America he credited Brady and his daguerreotype for his victory.[24] The picture Brady took that day became one of the most iconic photographs of all time.

Photographic studios sprung up like weeds and prices fell rapidly. Fancy studios in the big city centres offered elaborate props, potted palms and lush drapes. Less ambitious operators established premises in smaller towns and cities, and itinerant photographers travelled from town to village to town, setting up shop at county fairs, crossroads, or wherever they thought they might find a few people who would spend a day's wages to be immortalised.

Nantucket, USA, 1841

Frederick Douglass was born a slave in Maryland around 1818 – as he said in his autobiography, 'I do not remember to have ever met a slave who could tell of his birthday'[25] – to a slave mother and an unknown white father. Against the wishes of his owners, he taught himself to read and was particularly inspired by a schoolbook elucidating the classical principles of rhetoric.[26] When Douglass was about 20, he escaped and fled to the northern states, where he found work and became involved with the abolitionist movement. At an anti-slavery convention in Nantucket in August 1841 he 'felt strongly moved to speak' to the mainly white audience.[27] His speech was so successful that Douglass was invited to be the Massachusetts agent for the movement, whereupon he began touring the country speaking and writing against slavery.

From the moment he saw a daguerreotype portrait, Douglass realised it could be a powerful weapon in the battle against slavery. He understood that the new medium could change the way people saw the world, and specifically how they saw black people. Handmade depictions of black people in the United States at the time typically tended either toward an overtly racist, negative caricature, or the more sympathetic but equally racist idea of the noble but abused savage.

As early as 1841, Douglass used his own daguerreotyped image to change the way the white American public saw black people. As he travelled around the country campaigning and lecturing, Douglass made sure to have his daguerreotype taken as often as possible in one of the portrait studios that were springing up everywhere. His portraits revealed a novel image of an African American. They showed a handsome, elegant and intelligent man, immaculately dressed with head held high and eyes looking directly and intently at the camera. This man was not a curiosity or an object of pity on display but a human being: a man presenting himself on his own terms, and on equal terms with the viewer.

Photographic images of Douglass were turned into engravings and printed in pamphlets, books and newspapers. Douglass gave portraits as gifts to friends and admirers, and it was not uncommon to see his portrait hung in homes and offices.[28] When he launched his own newspaper, he used photographs of himself as giveaways to entice new subscribers.[29]

Douglass had a deep understanding of the power of the image to inspire change. He recognised its ability to persuade and alter perceptions, writing in 1861: 'As to the moral and social influence of pictures, it would hardly be extravagant to say of it, what Moore has said of ballads, give me the making of a nation's ballads and I care not who has the making of its laws.'[30]

There were at least 168 different photographic images of Douglass taken over his lifetime. This was more than Abraham Lincoln (126), General George Custer (155) or Walt Whitman (127), and made Douglass the most photographed American of the nineteenth century.[31] *

---

* The British Royal Family were even more enthusiastic early adopters of photography, with a reported 655 known images of the Prince of Wales, 428 of Queen Victoria and 676 separate photographs of Princess Alexandra (Stauffer et al., 2015, p.xii).

Beyond the Daguerreotype

Each daguerreotype was a unique image. It was produced on a shiny silver plate and sealed in a protective case to prevent tarnishing. Daguerreotypes were exquisitely beautiful but they were fragile and couldn't be reproduced, and their popularity declined when the successor to Talbot's negative–positive calotype process, Frederick Scott Archer's collodion method, was introduced around 1850. Talbot and Archer's images could be printed onto paper and produce multiple prints from a single negative, making them much cheaper and more convenient than daguerreotypes.

In 1854 a Frenchman patented a multi-lens camera that produced up to eight identical images at a time. He divided up the images and stuck them onto a small stiff card he called a *carte de visite*. These became a craze around the world. Individuals had their own *cartes* made to exchange as gifts with friends and family, and studios began selling *cartes* of the famous people of the day for people to collect, creating an early version of celebrity culture. And predictably, it didn't take long before a brisk under-the-counter trade in plain-wrapped pornographic *cartes* developed, along with equally predictable fears for society's imminent collapse.[32]

Beyond Portraits

Portraits are estimated to have accounted for 90 per cent of all photographs produced in America in the first fifty years of the medium.[33] Meanwhile, photography delivered on its promise in other fields. A couple of years after the daguerreotype, the world's first illustrated magazines were launched, the *Illustrated London News* in 1842, and *Gleason's Pictorial* in Boston and

Frank Leslie's Illustrated Newspaper in 1855. While these publications could not print actual photographs – this was not possible until 1880 – they published engravings made from photographs and described them as photographs, and they were generally accepted by readers as such. These magazines had huge circulations and marked the beginning of photo-journalism and of mass celebrity culture.

Roger Fenton photographed the Crimean War in the 1850s from a studio inside a horse-drawn caravan. One of Fenton's most famous photographs shows a deserted valley strewn with Russian cannonballs. In fact, there are two versions of this photograph: one with cannonballs on the road running through the valley and one with the road cleared. It remains a matter of debate whether Fenton staged the photo by scattering the cannonballs on the road to make it look more dramatic or whether the balls were taken away by soldiers in the course of the fighting.[34] Whatever the truth of the matter, it is fascinating how little time it took for nature's pencil – as Talbot described photography – to be accused of manipulation by human operators.

By the time the American Civil War began in 1861 the potential of photography to deliver a particular message was being exploited to the full. Matthew Brady assembled a mobile studio and darkroom and a team of assistants to document the war and its participants. His team produced unprecedented images of dead and wounded soldiers, bodies thrown into mass graves, and devastated battlefields. Here too, there are ongoing debates about whether Brady's and his colleagues staged some of their images for greater emotional or visual impact, and what difference it should make to our understanding of those images if they did.

In science it was the potential of photography to produce images the eye couldn't see that most excited practitioners. From the outset, photography was combined with microscopes and telescopes to extend their powers and produce images that could be studied, copied and shared with other scientists. In astronomy, long exposure times revealed details of distant stars that the eye alone could never detect, even with the most powerful telescope. Images of the Sun that would blind a human eye gave physicists new insights into the nature of our home star. Photographers such as Eadweard Muybridge and Étienne-Jules Marey found ways to take multiple rapid images of animals and people in motion that changed the understanding of movement and anatomy.

Along with the benefits photography brought science, there were misfires. In the late 1800s photography gave renewed impetus to the dubious study of physiognomy, a quasi-science that attempted to link facial features to character traits. The English statistician and social scientist Francis Galton developed a technique of composite photography to create 'average' faces, from which he tried to identify facial 'deviances' common to murderers and thieves. He couldn't, funnily enough. Another quasi-science, optography, was supposedly the practice of retrieving the image of the last thing seen from the eye's retina. In at least one case police photographed a murder victim's retina hoping to see the face of their killer revealed there (spoiler alert: it wasn't).[35] In another case, however, an accused murderer confessed when police told him they had retrieved his image from the dead victim's eyes.[36]

Photography and Art

When photography was introduced, commentators and critics expressed different opinions on its likely impact on art. Some expressed fears for art's imminent downfall.[37] Others felt photography would be a valuable tool for artists, as it would save them the time and trouble of studying their subjects from life, apparently failing to realise that this was like an eighteenth-century seamstress celebrating the invention of the sewing machine. Many, including Daguerre and Talbot, believed photography would stimulate interest in art by making images of the world's great works available to those unable to visit them in person.

All these predictions eventually came true to a greater or lesser degree. Photography did indeed bring about a steep decline in the demand for drawing and painting as a visual record. Portraiture, topographical and scientific illustration all declined in the decades after photography was introduced. Some of the jobbing artists whose livelihood had dried up became photographers themselves and earned a decent living cashing in on the portraiture boom.[38] Artists who were talented or established enough to carry on as painters often did use photographs as tools, although they were usually sheepish about admitting it.

In photography's early years nude photographs were officially registered with the French Academy as 'aides for artists' to avoid censorship (although, as we've seen, it didn't take long for pornographic images to become widely available on the black market). Later, enterprising dealers created whole catalogues of photographic images intended as artists' aides. Pioneering contemporary artists including Gustave Courbet and Eugene Delacroix were unapologetic in their use of photographs for reference and inspiration, though they faced criticism for it at the time.[39]

Meanwhile, art institutions did take advantage of the new medium to increase their reach and influence. The British Museum, the Louvre, Antwerp Museum and South Kensington Museum – precursor to the Victoria and Albert – all commissioned well-known photographers to record their collections, and created and published albums for sale to the public.[40]

Successful artists also took advantage of photography to create records of their own work, and in some cases were able to generate a secondary income selling photographs of their paintings. In reverse, the Trustees of the National Gallery of Victoria commissioned photographs of European works of art that would never otherwise be seen in Australia and displayed them for public view.[41]

Photography as Art

Far more controversial than these applications of photography to art was the notion that photography could be an art form in its own right. Talbot explored the artistic potential of photography from his earliest experiments, and described photography as 'drawing with light' and 'nature's pencil'. Other enthusiasts soon followed suit, much to the disgust of some critics. French poet Charles Baudelaire railed against the 'sun-worshippers' who dared to equate photography with art. According to him, photography was the refuge of painters 'too ill-endowed or too lazy' to complete their studies.[42] The French photographer Nadar was widely mocked for attempting to elevate his photographic portraits beyond pastiche, while British photographer Julia Margaret Cameron was called 'slovenly' when she experimented with effects such as soft focus.

It is true that some early experiments in artistic photography ranged from bizarre to downright silly. Oscar Gustave Rejlander's 1857 photograph, *The Two Ways of Life*, was modelled on a classical allegory and depicted a patriarch offering two youths the choice between sin and virtue. As a montage of thirty-two different photographs it was a technical masterpiece, but its subject matter and style look somewhat uncomfortable in the photographic medium, to say the least.

Photographic portrayals of nature were more critically successful. Carleton E. Watkins was the first photographer to
capture the grandeur of California's Yosemite with its magnificent granite cathedrals, domes, spires, skulls and hulking
monstrous shapes. At a time when it wasn't possible to enlarge
negatives, he mastered the technical and practical complexi-
ties of producing 'mammoth plate' images with an enormous
camera and glass plates that were bigger than a car window.[43]
A critic of the day called his photographs 'perfection of art' and
his work prompted President Lincoln to pass a law protecting
Yosemite and ultimately led to the establishment of the US
National Parks Service.

In Europe, Gustave le Gray's 1850s seascapes were virtuoso
examples of photography's potential both artistically and
technically. He caught the action of waves in photographs at
a time when the long exposure required made this extremely
difficult. He solved the problem of the different exposure times
needed to capture sea and sky by photographing them separately
and combining them in the print, joined along the horizon line.
His aim was artistic rather than documentary – some of his
images even combined sea and sky photographed in different
places – but at the same time the images he produced looked
closer to how our eyes see these natural features and so were
arguably more 'truthful' than a single camera image could be.[44]

The ambivalence over photography as an artistic medium in
its own right continued well into the twentieth century. Ernst
Gombrich's seminal book *The Story of Art*, originally written in
1950, contained only passing references to photography and no
examples of photographic art in his plates, though ironically all
the plates were photographs of (non-photographic) works of
art. It wasn't until the book's fifteenth edition, published forty
years after the original, that a 1952 photographic streetscape
by Henri Cartier-Bresson was included as plate number 404.[45]

And while there is no question today about photography's claim to be an artistic medium, even at the highest level it remains, literally, a poor relation to painting. The most expensive photograph ever sold at auction fetched a little over $4 million (a private sale reputedly fetched $2 million more than this but this is unverified), while the highest price ever paid for a painting, Leonardo da Vinci's *Salvator Mundi*, at $450 million, is more than 100 times that amount.[46]

In the end, it was photography's indirect effect on art that made the biggest impact.[47] As some predicted, photography usurped the role of the artist as the everyday recorder of images, but in doing so it opened the way for artists to explore more idiosyncratic ways of representing the world. Just as the printing press freed medieval scholars from the labour of copying and memorising manuscripts four centuries earlier, so photography freed artists from the requirement of verisimilitude.

Photography appeared at a time when artists were starting to experiment with new ways of representing what the eye saw in paint. Constable and Turner painted fleeting moments in time rather than idealised versions of reality in the early decades of the nineteenth century. Then around the time Daguerre launched his invention, a new artistic movement emerged in France, led by rebellious painter Gustave Courbet and the resolute 'peasant' Jean-François Millet. Influenced by Turner and Constable, these artists – who became known as Realists – depicted ordinary people and apparently ugly or banal features of the landscape with a care and scale that were traditionally reserved for much grander subjects. Others took up Courbet's mantle of Realism and focused on capturing images exactly as they saw them, rather than as they 'ought' to be, whatever they were. Led by Eduard Manet, they left their perfectly lit studios and explored ways of capturing the interplay of light and nature as it appeared outside, unfiltered. At their first exhibition in 1863, the now famous Salon des Refuses – established after their work was rejected by the official Paris Salon – scandalised visitors and critics with the style and subjects of the new works, such as Manet's *Dejeuner Sur l'Herbe*. One critic derisively dubbed Manet's successors '*les impressionists*', and the rest is art history.

Worth a Thousand Words?

Photography appeared two years after the coronation of Queen Victoria, and came of age alongside the telegraph, the transatlantic steamship and the railway. Together these technologies sped up time and shrank the world. Messages could be sent across continents and even oceans in minutes, and people and things could move around the world as never before. And they did. Nearly 3 million European immigrants arrived in the USA between 1840 and 1850, with many millions more to come in subsequent decades, and millions more Europeans moved to new colonies in South America, Australia, South Africa and New Zealand. Within the USA, tens of thousands moved from the cities of the east coast to newly available farmland in the Midwest, while in Europe industrialisation and crop failures were driving people from the countryside into growing cities with their promise of work in large, gaslit factories.

Human dislocation was taking place on a massive scale. Did photography play a part in this? Photographs showed people what was at the other end of the line, and what they'd left behind. Loved ones could be captured on a little plate, packaged in a frame and kept close. Did this make it easier to leave them behind? Photographs created the illusion of presence: a 'true' likeness, rendered by nature. The likeness was a mirage of course, devoid of a body's warmth, touch and characteristic smell, the sounds they made moving around and, most importantly, their voice. But the picture was something tangible, miraculous and, most importantly, present. We can be apart because we see each other.

And as photographs stood in for absent loved ones, the photographable – that is to say, visible – aspects of a person's being assumed a new significance, over and above their other characteristics. The memory of the sound of a loved one's voice, the grasp of their hand and the smell of their neck may fade over time, but the photographed face is never forgotten, even if that face is frozen in time.

Photographs captured what appeared to be an unmediated reality and were therefore assumed by many to be infallible envoys of truth. In an 1859 melodrama called *The Octoroon* one character, wielding photographic equipment, tells another:

The apparatus can't mistake. When I travelled round with this machine, the homely folks used to sing out, 'Hillo, mister, this ain't like me!'
'Ma'am,' says I, 'the apparatus can't mistake.'
'But, mister, that ain't my nose.'
'Ma'am, your nose drawed it.'[48]

This little exchange is the earliest known reference to the camera's supposed inability to obscure the truth. To this day there is a general belief that 'the camera doesn't lie', despite the fact that photographs haven't always delivered on their truthful promise. From the outset photographs were faked and staged to boost a particular cause or increase their dramatic effect. An 1895 newspaper article quipped: 'Photographers, especially amateur photographers, will tell you that the camera cannot lie. This only proves that photographers, especially amateur photographers, can.'[49]

Notwithstanding this sort of sophisticated scepticism, for most people, most of the time, photographs were considered accurate representations of reality and as such defined what 'is'. As portraits they allowed ordinary people to immortalise the best version of themselves they could muster. As personal mementos they created and fixed memories and transported loved ones through time and space. In journalism, politics and advertising – all historically dominated by written or spoken words – the role of photographs evolved from illustrating text to verifying it, abbreviating it and, finally, superseding it altogether.

In 1911 a journalist, writing an article about journalism, used the phrase 'a picture is worth a thousand words'. A decade later an ad man, writing an article about advertising, upped it to 'one picture is worth ten thousand words', and added the entirely fictitious claim that the phrase was a Chinese proverb, hoping this would make people take it seriously.[50] Given the relentless rise of images in advertising and elsewhere to their present ubiquity, and the general acceptance of the phrase as a truism, his strategy worked.

Henry Fox Talbot called photography a little bit of magic, realised. Humanity remains deeply under its spell.

13

SURPASSING IMAGINATION: MOVING IMAGES

Reality surpasses imagination;
and we see, breathing, brightening, and moving before our eyes
sights dearer to our hearts than any we ever beheld
in the land of sleep.

Anonymous, *Lights and Shadows of Scottish Life*, 1822

Georgia, USA, 1915

On 6 December 1915, of group of men strode down Peachtree Street, Atlanta, wearing white robes and pointy white hats with crudely cut eye holes pulled down over their faces, firing guns into the air as they went. The week before, a soldier-turned-preacher-turned-salesman called William J. Simmons had led a dozen of them to the top of nearby Stone Mountain. There, in an elaborate ceremony involving a makeshift stone altar, a Bible, an American flag and a giant, burning wooden cross, Simmons had declared himself 'Imperial Wizard of the Invisible Empire of the Knights of the Ku Klux Klan'. With that, the Ku Klux Klan was reborn for the twentieth century.

The original incarnation of the Ku Klux Klan was very short lived. It was formed in the late 1860s after the defeat of the Southern states in America's Civil War and during the conflict and confusion of the subsequent Reconstruction period. It was quashed by federal laws passed by President Ulysses S. Grant in the early 1870s, and by the turn of the century it was all but forgotten. Simmons had long dreamed of reviving it, and saw his opportunity in 1915 when the Klan exploded back into the national consciousness in no uncertain terms.

In February that year, the novel entertainment form known as motion pictures (or 'movies') had produced something that stunned everyone who saw it. A silent epic film called *Birth of a Nation* was more than three hours long and boasted a cast of thousands, with accompaniment by a forty-piece orchestra.[1] The film claimed to be historically accurate and told a tale of life during and after the Civil War, focusing on the fortunes of two families from either side of the conflict. In the film, newly freed black men and corrupt northern carpetbaggers ran amok in post-Civil War South Carolina, looting property, raping white women and feasting on government handouts. The politicians portrayed were unable or unwilling to step in. The chaos was going from bad to worse until, as a subtitle announced, 'at last there had sprung into existence a great KKK, a veritable empire of the South, to protect the Southern country'.[2]

The hero of *Birth of a Nation* was the Ku Klux Klan.

The film premiered in Los Angeles and made its way across the nation in the subsequent months, playing to thousands and earning millions of dollars at the box office. In December 1915 it was finally due to premiere in Atlanta, and it was there that Simmons and his fellow newly minted Klansmen were headed.

It turned out the Atlanta group was just the beginning. Fifty million people saw *Birth of a Nation* over the next five years, at regular screenings and special recruiting events sponsored by the KKK.[3] By 1925 the revived Ku Klux Klan had millions of members and in August that year 50,000 Klansmen and women in white robes and pointy hats marched with uncovered faces through the streets of Washington DC.[4]

Europe, 1600–1850

In seventeenth-century Europe the invention of the telescope and microscope prompted a great enthusiasm for optical devices. A popular entertainment was the 'Magic Lantern' spectacle, usually featuring angels, devils and other magical creatures painted on glass, illuminated by lamp or candlelight and enhanced by mirrors and lenses.[5] As these shows grew more ambitious, they experimented with ways to animate their characters so that they appeared to move.

Numerous optical devices followed that created the illusion of moving images. There was the zoetrope, the phenakistoscope, the zoopraxiscope, praxinoscope, motoroscope, lamposcope and tachyscope, among others. They all exploited a peculiar feature of vision, that the retina retains an image for a fraction of a second after seeing it. This phenomenon, known as persistence of vision, is why a series of slightly changing static images gives the impression of movement. You may have made 'animations' by drawing stick figures on page corners and flicking through the pages to make the figures 'move'. This is essentially the basis of all motion-picture technologies. The movement illusion kicks in at around fifteen images, or frames, per second; fewer than that and we see the individual images or very jerky movement (modern films are shot at twenty-four frames per second). The momentary flickering is critical to the movement illusion, as it disguises the changeover from one image to the next – running a film straight through a projector without flicker displays a blur. It took many years for movie technology to get to a point where the flicker was invisible – this kicks in when the rate of flicker passes the human 'flicker fusion threshold' of around fifty images per second* – hence the old-fashioned term for movies: the flicks.

Europe and USA, Late 1800s

By the second half of the nineteenth century several players around the world were competing to develop a method to make and display moving photographic images. The challenge was twofold: to devise a camera that could capture images in quick succession, and a device to animate and view the result. The critical requirement was a mechanism to achieve the stop-start movement or intermittent motion necessary for both capturing and displaying a sequence of images in very quick succession. In 1882 Étienne-Jules Marey adapted a revolving gun into a camera that captured several images per second (his predecessor Muybridge had used banks of separate cameras triggered by strings to capture his famous horse and human movement sequences). Around 1888 another Frenchman named Louis Le Prince invented a camera that could take several frames per second and recorded what is believed to be the first filmed sequence – a couple of seconds'

---

* Incidentally this may be why dogs don't often pay attention to TV. Their flicker fusion threshold is much higher than humans' (they are better at detecting motion) and, until high-definition TVs became available a few years ago, they would have seen TV screens – flicking away at 60 fps – as an annoying flickering light.

worth of his family perambulating in Roundhay Gardens in Leeds. He disappeared in mysterious circumstances before his technology was made public.*

Meanwhile, in the same year yet another Frenchman, Charles-Émile Reynaud, patented the *Théâtre Optique*. His system projected animated drawings using long strips of images passing from reel to reel. A sprocket mechanism engaged with perforations made along the edges of the strips to make sure each frame stopped – momentarily – in exactly the right place. While Reynaud's device projected drawings rather than photographic images, his sprocket mechanism made a crucial contribution to cinema technology.

The following year, Thomas Edison applied for US patents for the Kinetograph and the Kinetoscope, a motion picture camera and moving picture display unit, respectively. These machines combined elements of earlier inventions – rolls of perforated photographic film moving from reel to reel, a rotating disc shutter and a stop-start mechanism – with recently invented celluloid film and other improvements to create the world's first moving picture displays. The Kinetoscope wasn't a projector but a peephole unit and Kinetoscope parlours had banks of individual viewing machines showing short filmed sequences. The first one opened in New York in 1894 and sites soon followed in other US cities and Europe.

---

* Louis Le Prince vanished without trace in 1890 while taking a train from Dijon to Paris. His body was never found. His son, Adolphe, later sued Thomas Edison over their respective cinematic inventions, and eventually won. Adolphe was shot dead while duck hunting in New York in 1902, and no one was ever tried for the murder. Some believe that Thomas Edison was behind both incidents. A note handwritten by Edison days after Le Prince Senior disappeared says, 'It has been done. Prince is no more', and goes on to mention his discomfort at the idea of 'murder'. (Gupta, A., 'The disappearance of Louis le Prince', *Materials Today*, Vol. 11, Iss. 7–8, p.11, accessed at www.sciencedirect. com/science/article/pii/S1369702108701603.)

Back in France, a pair of brothers was working on their own moving-picture system. Auguste and Louis Lumière devised a combined film camera, printer and projector they called a Cinematographe. It used an adapted sewing machine mechanism to create the intermittent movement, and a revolving shutter to create the flicker. Their device was almost 100 times lighter than the Edison machines, and mechanically simpler, making it much easier to transport, maintain and use.[6]

On the afternoon of 28 December 1895, a small crowd of people paid a franc each to sit on rows of folding chairs and see an exhibition of the Lumière brothers' filmed *actualites* in a small room in the basement of the Grand Café on the boulevard des Capuchins in Paris. Surrounded by potted palms and dodging one another's hats, the world's first paying cinema audience saw ten silent films, none more than a minute long. They included the *Workers Leaving the Lumière Factory in Lyon*, a slapstick comedy sketch (*The Gardener*), scenes with the filmmaker's baby, and children diving from a jetty into the sea. Word of the amazing new spectacle spread – imagine the amazement of the time: pictures of real people and things moving before your very eyes! – and within a few weeks policemen were managing the lines of people waiting for admission, and the shows were making thousands of francs per day.[7] Over the next year dozens of new 'views' (as the films were called) were added, such as the famous *Arrival of a Train at La Ciotat*. According to legend, this film caused audiences to flee from their seats for fear of being crushed by the oncoming train. The Lumières built 200 cinematographe machines and trained a fleet of operators to take the show around the world, filming new views as they went. For a couple of years they were wildly successful, but they were also widely copied, and when their films failed to evolve, their audience numbers began to fall. It would be for others to take the new art of cinema into the twentieth century.

USA, Early Twentieth Century

It didn't take long for Thomas Edison and other entrepreneurs interested in the new medium to imitate the Lumières' innovation, and various film projection systems appeared after 1895. Shown in dedicated stores or in theatres as part of a vaudeville programme, the films were short slices of life, comedic sketches or magic shows playing around with the 'tricks' cinema could perform, such as making people disappear and reappear. For the most part they were a technological novelty. Stories were basic or non-existent. Cameras were generally static and action usually took place in a single take.

From 1905 dedicated shopfront cinemas known as 'Nickelodeons' began to appear in shopping precincts and urban neighborhoods charging 5 cents (a nickel) admission for programmes lasting around fifteen minutes. By 1910 there were more than 10,000 Nickelodeons across the USA, and a third of the nation went to the movies every week.[8] As directors on both sides of the Atlantic experimented with longer, more narrative-driven films, entrepreneurs built dedicated movie theatres to attract bigger, more middle-class audiences.

Hollywood, 1915

In 1908 an actor and would-be playwright named David W. Griffith started making films for the Biograph company including, as it happens, the very first film made in Hollywood (*Old California*, 1910). In 1914 he optioned a novel called *The Clansman* written by Thomas Dixon Jr: the story of the Ku Klux Klan's 'heroic' role in stabilising the South in the chaos of post-Civil War Reconstruction. Dixon was a Southerner of the old school, the son of a slave owner and member of the original

KKK. His stated objective was 'to demonstrate to the world that the white man must and shall be supreme'.[9] Nine months later, D.W. Griffiths released a three-hour-long silent movie based on Dixon's novel, now renamed *Birth of a Nation*.

The film caused a sensation. Full-length films with huge casts and extravagant sets had been made before, but this was something completely new. *Birth of a Nation* was uniquely powerful because Griffith made it using filmmaking techniques that had never been seen before. He varied the distance between audiences and the screen by using different types of shot: long, full, medium and, famously, close-up. Close-ups had been used occasionally before but they were still a novel idea. The film's star Lilian Gish said years later that not everyone working on the film got the point of them:

> The people in the front office got very upset. They came down and said: 'The public doesn't pay for the head or the arms or the shoulders of the actor. They want the whole body. Let's give them their money's worth.' Griffith stood very close to them and said: 'Can you see my feet?' When they said no, he replied: 'That's what I'm doing. I am using what the eyes can see.'[10]

Griffith moved his cameras, following the action from a fixed point, in panning shots and, even more radical, moving the whole camera alongside the action in tracking shots. These brought the audience into the action, giving them the impression they were riding alongside the Klansmen or following a running character as they might in life. He varied camera angles from the standard head-on approach and used lighting to create mood and guide the audience's gaze around a scene. He intercut shots from different scenes that were apparently happening in parallel – damsels in distress hiding in a basement, the bad guys looking for them, and the KKK rescue party on the way. As well as distorting time and space in a novel way, it created tension, as he made the cuts shorter and shorter as the climax approached.

Editing was critical to *Birth of a Nation*. The film had 1,544 shots at a time when films of similar length typically had around 100. The shots were edited together carefully to minimise their obtrusiveness and look as 'natural' as possible. Griffith would set a scene with a wide 'establishing' shot then cut to other perspectives to guide the audience's eye around. When two protagonists were squaring up to each other, for example, he would cut between them so the audience saw each one as the other protagonist would see them. This novel editing style is now called 'continuity' or 'Hollywood style' editing, and remains the approach most directors use to this day.

Crucially, and again uniquely at the time, Griffith harnessed all these techniques to serve the film's story. They were designed to manipulate the audience's emotional responses rather than to create a visual spectacle. This was a new experience for audiences, and they loved it.

Motives ranging from artistic ambition to white supremacist propaganda have been attributed to Griffith for the content of his film. What is not a matter of debate is the incredible persuasive power it had. On seeing the film, *The Clansmen* author Thomas Dixon Jr described cinema as 'the mightiest engine for moulding public opinion in the history of the world'.[11]

Dixon's motives may have been sinister but his analysis was correct. *Birth of a Nation* revealed the power of motion pictures to move audiences and, in so doing, to persuade them. With apparently effortless ease, it demonstrated that film could convey an argument with a force the most powerful rhetorician would envy. Griffith was the first filmmaker to realise that the motion picture medium, skilfully deployed, 'could exercise enormous persuasive power over an audience, or even a nation, without recourse to print or human speech'.[12] Griffith, more than any other film pioneer, established film as a new medium of communication and expression with its own, uniquely visual, grammar.

Moving pictures combine the power of the image to simulate reality and stimulate the imagination with the power of the written word to direct an audience's flow of attention. Unlike the written word, they require no tuition and very little effort to absorb. They can convey some ideas far more quickly than words – the expression in a glance for example – and in subtle ways of which the audience may not even be aware.

Moscow, USSR, 1919

The persuasive potential of the new visual 'language of film' was not lost on the fledgling Soviet government in post-revolutionary Russia. The Bolsheviks, led by Vladimir Lenin, were desperate to knit together the enormous, disparate and widely illiterate Russian population after years of civil war. They formed a Cinema Committee within the Department of Education, led by Lenin's wife, Nadezhda, nationalised the entire Russian film industry under its umbrella, and established the Moscow Film School to teach Soviets how to make pro-Bolshevik propaganda films, known as Agitprop.[13] *

Unlike American filmmakers preoccupied with a film's potential box office, the Moscow Film School faculty were free to explore the more theoretical aspects of filmmaking. They were interested in its psychological effects, especially its persuasive power, and in how editing techniques could create different narratives and emotional effects.

---

* Agitation and propaganda.

One of the Film School's founders, Lev Kuleshov, conducted a series of experiments in which he showed the subjects three film sequences: a well-known actor's face followed by a plate of food; the same actor's face then a woman in a coffin; and the actor again, this time followed by a scantily clad woman posing on a sofa. Subjects praised the actor's skill at portraying, in turn, hunger, grief and lust on screen. In fact, as you may have guessed, each of the three shots of the actor were identical.

Kuleshov established systematically what Griffith had discovered instinctively. People naturally link shots that are joined together sequentially regardless of whether, taken individually, they bear any relation to one another. He realised that, in film, the juxtaposition of shots – in other words, editing – was at least as important as the content of the individual shots.[14] This was film's unique capability, he concluded, and this insight became the basis of a new cinematic technique called montage that involved editing sequences of unrelated images together to create new meanings in the viewer's mind.* Kuleshov and his colleagues demonstrated conclusively that motion pictures spoke to audiences in a whole new way a decade before the 1927 addition of sound.

The study of montage culminated in Sergei Eisenstein's 1925 pro-Bolshevik propaganda masterpiece *Battleship Potemkin*. It was composed of more than 1,300 shots – none more than a few seconds long – edited together in rapid succession to tell the story of a real-life 1905 uprising of sailors in the Tsarist navy and the subsequent massacre of civilians on the Odessa Steps. The film was banned in the UK, USA, France and Germany for fear it would stir up revolutionary feelings. In the UK it was

---

* After a century of modernist and experimental art forms that play with the juxtaposition of apparently unrelated ideas, this doesn't seem radical now, but it was seen as highly original at the time.

only released for general viewing in the 1950s, when audiences for silent films were – safely – almost non-existent.

Nazi Germany, 1933

One of *Potemkin*'s admirers was the Nazi propaganda minister Joseph Goebbels. As he told an audience of German filmmakers, 'Anyone who had no firm political conviction could become a Bolshevik after seeing the film'.[15]

It didn't take long for the Nazis to follow the Bolsheviks' example. A 1934 order proclaimed: 'Hereafter the State will assume complete responsibility for the creation of films. Only by intensive advice and supervision can films running contrary to the spirit of the times be kept off the screen.'[16]

In 1932 a young actress and filmmaker called Leni Riefenstahl wrote an admiring letter to Adolf Hitler. Riefenstahl was famous for her appearances in so-called 'mountain films' – a peculiarly German genre much favoured by right-leaning audiences for their portrayals of the Teutonic ideal – and had directed one of her own. Hitler was a movie fan, and Riefenstahl became part of the Nazi social milieu. Hitler invited her to film the rally planned for Nuremburg the following summer, putting almost unlimited resources at her disposal.

While Riefenstahl maintained forever after that her feature-length film, *The Triumph of the Will*, was simply reportage of the 1934 Nuremberg event, it was a production on a scale worthy of any Hollywood epic. She worked alongside the Nazi architect Albert Speer to design the event for maximum cinematic impact, as she described in a 1935 interview (though later denied): 'The ceremonies and precise plans of the parades, marches, processions, the architecture of the halls and stadium were designed for the convenience of the cameras.'[17]

The stages had pits dug in front of them for cameras, dolly tracks were laid around and through the parade grounds, and a giant elevator stood beside the main podium to capture soaring views of the crowd. Riefenstahl directed a crew of 170 men operating lights, cameras, huge cranes and dollies, and bulky sound equipment to record the event as dramatically as possible. More than half a million 'extras' attended the rally, including thousands of exhaustively rehearsed soldiers, and clean-cut blond civilians in traditional German dress. The medieval city's towers and spires formed a gorgeous backdrop, and countless fires, banners, guns and flags served as props. All that was topped by a mesmerisingly charismatic leading man as the undisputed star of the show. Where reality didn't measure up, the magic of cinema filled the gaps: certain scenes were reshot after the event in a studio, and events were edited out of chronological order where this increased the drama.[18]

The Triumph of the Will was a triumph of the art of filmmaking, and of propaganda. Hitler was presented as a mythical, godlike figure, swooping down from the clouds in a plane in the opening scenes to greet the waiting masses. Once he emerged from the plane, he was shot throughout from simperingly low angles, his head crowned by clear blue German skies as he appeared to tower over increasingly rapturous crowds.[19] There were visual references to ancient German mythology and to a more recently glorious past, with the implied promise that Hitler would restore the nation to a state of deserved, divine grace.

Hindsight gives the film a distinctly chilling potency, but at first, Western powers underestimated its devastatingly sinister impact. Riefenstahl was awarded a medal for the film at the prestigious Paris International Exposition and her next film, the equally potent propaganda film *Olympia* 'documenting' the 1936 Berlin Olympics, won the Gold Medal at the Venice Film Festival.

London, 1939

It wasn't only white supremacists and totalitarian regimes that weaponised motion pictures. When Britain declared war on Germany in 1939, the British Ministry of Information wasted no time in establishing its own Film Unit to make films of 'authenticity and optimism to inspire a nation'.[20] With more than a billion cinema admissions in the UK in 1940,[21] cinema-going was by now the primary form of entertainment, with more than two-thirds of people going to the pictures at least once a week.[22] They went to be entertained but also to see newsreels and documentaries. The MOI Film Unit produced both features and documentaries for the film-hungry wartime public. One short film set scenes from *Triumph of the Will* to a popular dancehall song to ridicule Hitler and the Nazis. Dylan Thomas also co-opted *Triumph* footage into a more sombre tirade called *These are the Men* in 1943.

Washington DC, 1941

By the time the USA joined the war in 1941, the Hollywood studio system had been in full swing for two decades and cinema was by far the dominant form of mass entertainment. It was also many people's main source of news. Several leading directors of the day – including John Ford, Frank Capra and John Huston – left Hollywood and enlisted to make films in support of the war effort. When Capra saw the banned Nazi film he despaired. In his later autobiography he remarked that the *Triumph of the Will* 'fired no gun, dropped no bombs. But as a psychological weapon aimed at destroying the will to resist, it was just as lethal.'[23] Nevertheless, he co-opted excerpts from Riefenstahl's film in his own series of morale-boosting training films, *Why We Fight*. Like his British colleagues, he used the Nazi's own propaganda to illustrate the evil of their cause.

Present Day

In recent decades, neuroscience has shed new light on the power of film to engage and move us. In 1991, a team of Italian researchers at the University of Parma were investigating the role of motor neurons – the specific parts of the brain responsible for different types of movement. In the course of their work they wired up the brains of several macaque monkeys. One day, quite unexpectedly, they noticed that when a monkey saw a researcher reach for a peanut, the monkey's brain lit up in the same way as if it had performed the movement itself. The monkey's motor neurons 'mirrored' the observed movement of the researcher, even when the monkey was still.

The team had discovered 'mirror neurons' in the monkeys: a network of brain cells that fire in a similar way when the body is doing something *and* when the eyes see it done.[24] Subsequent research using fMRI technology has established that humans have a so-called mirror mechanism, too. When we see someone reaching for a mug or running for a bus, the areas of our brain that control hand and arm movements or running actions, respectively, are unconsciously activated. Some researchers believe the phenomenon extends to sensations and emotions, so that if we see someone experiencing pain or disgust, for example, the parts of our own brain that process those feelings also fire up.

The mirror mechanism is, self-evidently, not identical to a physical experience and, as is so often the case in matters scientific, scholars disagree about its extent, impact and role in human brains.[25] Nevertheless, according to Vittorio Gallese, one of the Parma scientists who discovered the monkeys' mirror neurons, the mirror mechanism may explain our love of film and television. When we watch moving pictures on a screen, he says, we experience our own physiological version of what is being played out before us. As we watch moving images, Gallese maintains, we are 'watching' with our whole bodies. He calls the phenomenon Embodied Simulation. When we watch a filmed sequence, we see the screen and hear the audio, but our response goes beyond the audio-visual: our mirror neurons translate the physical actions we see on screen – movement, drama, pain, pleasure – into our very bodies. This brain/body response can take place at an unconscious level; that is, our brains can react in a way that we aren't even consciously aware of. Thus, a filmed scene can trigger a deep, visceral response that the same scene presented in written form – which, by definition, must be processed consciously through reading – could never achieve.

Other recent research reveals strong evidence of similarities between perceiving scenes in movies and how we see in life.[26] The techniques of 'Hollywood style' editing of the type first developed by D.W. Griffith are consistent with the way humans direct their attention and how we experience space, time and action. In other words, film editing mirrors the way we see the world.

When we look at things we tend to alternate between taking 'wide' views of our surrounding environment and focusing on specific details within a scene, without noticing the transitions from one to another. In the same way an edited film sequence will typically start with an establishing shot and cut to medium wide shots and close-ups. We move as we look, just like cameras do, and our conscious vision often jumps from one scene to another quite different view – think of looking up from a page, or from your car's dashboard to the road ahead. This research suggests that our brains process filmed images easily because they mimic our natural visual system, and the elements of film map onto our normal perception of reality. A film's action is equivalent to what is going on around us, camera movements equate to our bodily movements, and editing is what we choose to attend to from moment to moment.[27]

In one recent study, researchers compared eye movements and brain responses when subjects watched an edited Hollywood film sequence (from *The Good, the Bad and the Ugly*) and an unedited scene of everyday life filmed with a fixed camera. When watching the 'real life' sequence, the subjects' gazes moved around the scene in many different directions. By contrast, when watching the Hollywood film, their gazes moved similarly from moment to moment, suggesting the director, using the tools of filmmaking, was tightly controlling their attention. When viewing the everyday scene, each subject's brain activity varied, showing just 5 per cent correlation from person to person. When they watched the Hollywood movie, however, brain activity was correlated by 45 per cent.[28] The Hollywood filmmakers were not only managing the audience's gaze, they were successfully controlling their minds.*

These studies go some way to answering the question posed by critic Steven Shaviro: how can cinema have such a powerful 'reality effect' when it is so manifestly unreal?[29] Filmmaking techniques, many of which have analogues in natural vision, give filmmakers the tools to exercise a high degree of control

---

* In a comparable study, a 1961 *Alfred Hitchcock Presents* television episode achieved an even higher rate of brain correlation among subjects – 65 per cent – providing quantitative evidence for Hitchcock's reputation as the 'master of suspense'.

over viewers' attention. This translates into viewers' emotions as they experience a physiological – as well as intellectual – response. The 'moving' aspect of motion pictures – the combination of action, camera movement, and editing – and the brain's mirror mechanism combine to take audiences, mind and body, *into* film in a way that no other medium has ever achieved.

1940s, Post-Second World War

In 1841, just a couple of years after Daguerre's famous announcement, a Scottish watchmaker called Alexander Bain submitted a patent application for an apparatus to scan images and transmit them electronically – the basis of television. Various similar inventions followed, but it wasn't until 1926 that the first television images were successfully transmitted by another Scotsman, John Logie Baird. The earliest TV systems were mechanical, had poor-quality pictures and required huge, unwieldy cameras. These were replaced with electronic systems using cathode rays in the 1930s, but the outbreak of the Second World War halted progress and television remained more or less dormant throughout.[30]

Television transmission and production resumed in the mid-1940s. The industry structure that countries adopted for television differed according to government ideologies and expectations of the new medium. British television was produced and broadcast exclusively by the publicly owned (but politically independent) British Broadcasting Service, already the monopoly provider of radio. Households with a television paid an annual TV licence fee, and this money paid for the BBC – a model that persists to this day. Television advertisements didn't appear in Britain until 1955, when the Independent Television Authority (later ITV) set up a network of regional companies to provide commercial television. In the US, television was always a predominantly commercial activity.* Privately owned stations were set up city by city, or rather, market by market. The national radio networks NBC, ABC and CBS began forming networks of television stations from the late 1940s and were soon established nationwide. These 'Big Three' networks dominated US television until the rise of cable television in the 1980s.

Early television borrowed heavily from its older siblings, radio and theatre, with sitcoms and variety programmes trans-ferring neatly onto television. There were news, quiz and chat programmes, and plays staged much as early films had been, with fixed cameras recording the action as if in a theatre. Short films and cartoons were recycled from the movies. Screens were tiny by today's standard – 8 to 10in across – and picture and sound quality varied. Images were black and white with dis-cernible flickering, and sets were large, hulking wooden boxes.

But how those hulking boxes and tiny flickering screens sold! A million sets were sold every two months in 1950s America. By the end of the decade, 46 million households – nearly 90 per cent of all homes – had sets.[31] In cash-strapped, still-rationed Britain, set sales were slower to take off, but with the televised Coronation of Queen Elizabeth in 1953 and the relaxation of consumer credit rules in 1954 they soon caught up. Two-thirds of British homes had televisions by 1960 and more than 90 per cent by that decade's end. More British homes had a television than had a telephone, refrigerator or indoor lavatory.[32] Other Commonwealth countries also embraced television: TV penetration in Australia and Canada was over 90 per cent by the mid-1960s.[33] In France and Germany, still recovering from the war, rollout was slower, but by the end of the 1960s two-thirds

---

* Some cities had publicly funded educational television stations, and the national Public Broadcasting Service began broadcasting in 1969.

of French and three-quarters of West German households also had televisions.[34]

From the outset it was apparent that television was going to have a major impact on society, though surely no one could have anticipated just how far reaching that would be. Despite their shortcomings, those small, flickering screens proved irresistible. As soon as a family acquired a television set their daily routines were transformed. For several hours a day, adults and children alike turned their chairs to face the large wooden box in the corner of the room, and sat obediently watching whatever appeared before them.

The TV Times

Once people had a television set, they watched it. Early transmissions were quite limited – just an hour and a half per day in Britain, for example, and a couple more in the USA – but as the number of hours broadcast increased, so did viewing time. Within a very short time, American families were watching ten hours of television a week, and this soon doubled. Individuals soon clocked up around five hours per day in front of the TV. Europeans always watched a little less – settling at around four hours each a day. Time spent watching television remains similar to this day.

This begs the question: what did people *not* do for four or five hours a day once they started watching television?

Among children, playing with friends was the biggest casualty,[35] and this went along with a decrease in the amount of time they spent outside. Cinema going, reading and doing 'nothing' also declined, but the biggest change television brought about in children's lives was the shift of after-school and weekend time from the outdoors into the living room.

Women with televisions spent less time doing evening house-work, while men and women both spent less time involved in activities outside the home. Civic and social organisations, asso-ciations such as lodges and societies, sporting leagues, craft and activity circles, parent–teacher groups, religious, educational and philanthropic organisations, and volunteering to lead activ-ities such as the Boy Scouts and Girl Guides all saw significant declines in participation in the decades after television. Less structured social activities such as visiting and hosting friends, and going out to pubs, dancehalls, cafes, cinemas and theatres also suffered.[36] The slow and steady disappearance of these institutions in the subsequent decades is well known. About 15,000 British pubs closed between 1969 and the mid-2000s when smoking bans further accelerated their decline.[37] France has fewer than a third as many cafes today as it had in 1960.[38]

Scholars, as ever, disagree about whether television was the *cause* of these changes. They did take place at the same time, and it is also the case that declines in participation in collective activities took place to a greater degree in homes with televi-sions than in homes without, and to a greater degree in homes that watched more rather than less TV.[39] It is also undeniable that television represented the main change in how people spent their time over the second half of the twentieth century.

All this goes to say that, while television may not have been acting alone, as people watched more and more television, they saw less and less of each other.

In 1996 Harvard sociologist Robert Putnam described these changes as indicative of an erosion of 'social capital' in the USA over the previous three or four decades. Putnam defines social capital as all the aspects of a society that turn 'I' into 'we'. Specifically, it includes the 'networks, norms and social trust' that make it easier for people to cooperate; it is part of the glue that holds societies together. Putnam argued social capital was in decline, both in its inputs – collective activities such as the ones described above – and its outputs: levels of civic engagement and trust within a society. He cites declines in voter turnout and falling levels of trust in government and in one another as evidence. The proportion of Americans who felt 'most people can be trusted' almost halved (from 58 to 37 per cent) between 1960 and 1993.[40]

Putnam acknowledged several alternative causes for these changes, including more working women, and changing family and economic structures, but concluded that by far the biggest factor disrupting social-capital formation was what he called the 'individualisation' of leisure time by technology. In other words, television.[41] *

Eyes on Television

One of the early fears people had was that television might damage children's eyesight. 'You'll go blind watching that thing' was a common refrain, along with entreaties to sit further away from the box for fear of electric radiation from the screen. There has been a steady, worldwide increase in myopia – short sightedness – in the decades since television appeared. Were our parents right after all? Not quite. Decades of eye research have established no link between television viewing and increased likelihood of myopia. However, recent studies have established there *is* a clear link between time spent outside and eyesight.[42] Children who spend more time outside – where, even on a dull day, light levels are at least ten times higher than

---

* Later research showed that while overall television viewing was indeed correlated with lower civic engagement, viewing of serious news programmes was correlated with higher levels of civic engagement.

the brightest artificial light – are less likely to become short sighted. One study found each additional hour spent outdoors a week reduces the odds of myopia by 2 per cent.[43] Thirty hours a week watching TV – the average for children – is a lot of hours not spent outside. I wish I'd known that when I rushed home from school to watch *Gilligan's Island* instead of riding my bike around the block with my friends.

What's On TV?

The rapid and permanent encroachment of television into people's habits and schedules in the mid-twentieth century prompts an even more pressing question. What were people actually *seeing* during all those hours watching TV?

While television always showed news and documentaries, the overwhelming majority of the time people spent in front of television was, and remains, watching entertainment programmes. Sitcoms, dramas, variety and chat, talent and quiz shows, lifestyle, makeover and 'reality' programmes have been the staple diet of television from its beginning. And alongside entertainment, in the majority of cases, went advertising.

Entertainment television – both the programmes and the advertising that accompanied it – provided viewers an endless parade of almost-attainable perfection, or at least aspiration: perfect teeth, attractive faces, luxuriant hair, shapely bodies, immaculate homes, the latest appliances and gadgets, happy families with a dad just home from the office in a suit, a mum (or mom) in an apron and two or three attractive children.

Television and cinema have always been quite different media. Television is by its nature smaller than cinema, and not just in terms of screen size. TV programme budgets were always much lower than films', so they used far less complex camera work and editing: millions of hours of television were recorded in real time using two or three fixed cameras. Where cinema was dazzling, television was more homely; it operated on a human scale. Its stars were a bit more approachable than the gods and goddess of the silver screen, its sets and backgrounds a little more accessible than Hollywood's epic vistas. TV lived in the intimate and familial environment of the home. It was *part* of the family.

All those hours and hours, seeing those friendly faces and places, sets and scenarios time and time again gave television a unique status … it may have been smaller in stature than cinema but television had at least an equally powerful, if more insidious, effect on how its viewers saw the world.

In TV Land ordinary people – just like you and me, but funnier and better looking – could have it all. Television showed people how to be – good citizens, good housewives, good men, good mothers, good boys and girls – or at least how to *appear* to be. Millions of people around the world bought into the ideal world presented on television. It appealed to the majority, who were happy to embrace an idealised reality after the upheaval and privations of the Second World War. As such, television both created and reflected the consumer society of the 1950s – a profound and lasting change in Western lives and economies.

Not everyone was included in the world they saw on television. Non-white faces were virtually absent from television for years (the blacked-up entertainers on the BBC's *Black and White Minstrel Show* (1958–78) don't count) and lifestyles outside the suburban family 'norm' were also either invisible or presented as caricature (Mr Humphreys, are you free?). This is to some extent a natural consequence of television's nature as a broadcast medium. When there was one family set and only a few channels, television inevitably catered to the majority: the path of least resistance into the living room was programming every member of the family could at least tolerate.

Even broadcasters such as the BBC, who didn't rely on advertising, was not immune to the pressures to maximise viewers, and the result was often bland and banal. With a few exceptions, minority interest television didn't appear until niche channels launched on cable and satellite networks decades later.

A Dramatised World …

There has been another more subtle, but far more profound, consequence of people watching all those hours and hours of fictional television programmes over many years.

While television entertainment may be less sophisticated than cinema, it operates on viewers' emotions in similar ways. When watching television – an inherently passive activity – images and sounds pour into viewers' eyes and ears, feeding their brains a stream of fictitious sensation. A whole catalogue of emotions might be elicited in a single night's viewing: fear, anger, sadness, joy, disgust, surprise, trust, anticipation, shame, envy, pity, love. Even when we know they are fictitious, the illusion persists.

Over time, television audiences came to expect a dose of emotion with every televisual meal and, ever fearful of a channel change, programme makers gave them what they wanted. The techniques used to make entertainment shows leaked into other types of programme. No genre has been immune to the pressure to add a 'feel along with me' aspect to every programme. Nature documentaries present viewers with 'characters' from the animal (and even plant) kingdom and exhort the audience to feel the trials and tribulations of life in the wild. History is no longer discussed by historians but increasingly dramatised with whatever licence the producers feel is necessary to keep the audience caring and enthralled. Even naturally dramatic events, like sport, need an emotional boost, so coverage of sporting events becomes as much about the players' personal journey or the fans' agony as about the final score. Emotion has found its way into what should be the most rational, and least fictional, genre of all: the news. Straight reporting of news is now an anachronism, as almost every news story incorporates an emotional angle or an individual human drama or, preferably, both.*

Over many years, these entertainment TV techniques have leaked beyond our screens and out into the real world. It started innocuously enough, when political, sporting and other public events started being organised to fit around television schedules. Then it crept beyond the timing and into the content of real world events. Today, drama and narrative have usurped rational analysis and abstract ideas. Politics, education, business – you name it. Everything has become infused with the techniques of entertainment television. Rationality has slowly but surely given way to drama and emotion in almost all walks of life.

Just as television viewers in the 1950s became the happy consumers they saw on TV, in later decades this effect mutated and spread: the real world became a version of the glossy replica portrayed on TV.

Chicago, September 1960

Six weeks before the 1960 presidential election, the incumbent vice president and Republican candidate, Richard Nixon, accepted his opponent's challenge of a series of televised debates, the first of their kind. Nixon was known to be a confident and adept debater. The week before the first debate, Nixon was in

---

* See the *Anchorman* films for a brilliant portrayal of this transformation in news coverage.

hospital with a knee injury and an infection. On the night of the debate, under the harsh TV studio lights Nixon appeared pallid, gaunt and sweaty. His grey suit sagged and blended into the studio background. The shadow of his beard showed dark under his pale skin.

Nixon's opponent, Massachusetts Senator John F. Kennedy, spent the morning of the first debate sitting in the sun and the afternoon napping. That night on air he looked tanned, young and healthy in his crisp white shirt and dark suit, despite being only four years' Nixon's junior and a lifelong sufferer of ill health himself.

The debate was watched by more than 70 million people – the highest television audience in history. A much smaller number listened to it on the radio. Radio listeners polled at the time gave Nixon a clear victory. Television audiences, however, favoured Kennedy, and he later credited television for having won him the presidency, just as Abraham Lincoln had credited an earlier visual technology for his victory exactly a century earlier.[44]

Television's future in politics was assured. Two decades after Kennedy and Nixon took to the podium on TV, the USA got its first film-star president. Another three and a half television-infused decades later, a reality television star took over the White House.

14

SEEING, WEAPONISED: SMARTPHONES

Look on my Works, ye Mighty, and despair
Nothing beside remains.
Percy Bysshe Shelly (1792–1822), 'Ozymandias', 1818

San Francisco, 2007

On a cold January morning in downtown San Francisco, several thousand people stood in line outside the enormous Moscone West convention centre. Some had been waiting all night. Apple CEO Steve Jobs was due to give his keynote address to the annual Macworld convention that morning and the faithful were there to hear it.

Ryan Block, star reporter for consumer technology website engadget.com, was in the crowd. He posted a live running commentary of the event:

> 8:56 a.m. – We're in! Almost the same seat as last time we did this.

8:57 a.m. – Everyone's shuffling in – including Jobs' family, who just rolled past. When the press barrier came down it was a mad rush to the elevator, people being trampled.

9:06 a.m. – Apple's standard tracks are playing: Gnarls Barkley, Coldplay, Gorillaz. You'd think they'd cycle through some of these tunes.

9:12 a.m. – 'Good morning ladies and gentlemen, Welcome to the Macworld 2007 keynote address.' Just a few minutes more now here.

9:14 a.m. – Ok, here we go. They're playing James Brown. Unreal applause for Steve – he's wearing the classic Jobsian garb. [Dark turtleneck tucked into jeans, white trainers.]

9:15 a.m. – People are standing on seats. 'We're going to make some history together today.'

Half an hour into the speech, Block was still transcribing and posting Jobs' words:

9:41 a.m. – 'This is a day I've been looking forward to for two and a half years. Every once in a while, a revolutionary product comes along that changes everything. One is very fortunate if you get to work on just one of these in your career. Apple has been very fortunate that it's been able to introduce a few of these into the world. In 1984 we introduced the Macintosh. It didn't just change Apple, it changed the whole industry. In 2001 we introduced the first iPod, and it didn't just change the way we all listened to music, it changed the entire music industry.'

9:42 a.m. – 'Well today, we're introducing THREE revolu-tionary new products. The first one is a widescreen iPod with touch controls.' The crowd goes wild. 'The second is a revolutionary new mobile phone.'

9:43 a.m. – 'And the third is a breakthrough internet communications device.' Tepid response on that last one, but he almost got a standing ovation on the phone.

'An iPod, a phone, an internet mobile communicator. An iPod, a phone, an internet mobile communicator … these are NOT three separate devices!'

A photograph appeared on the huge screen behind Jobs and the faithful jumped to their feet, cheering.

'We are calling it the iPhone!'[1]

The iPhone sold a million units in its first three months and more than 10 million units the following year.[2] Sales doubled in 2009 and again the following year, rising to a peak of 231 million units in 2015, before plateauing at around 215 million a year. In August 2018 Apple became the first company in history to be valued at over a trillion American dollars, based overwhelmingly on the success of the iPhone. In August 2020 it was worth double that amount.

Back at the 2007 Macworld convention, Google's CEO Eric Schmidt took to the stage to congratulate Steve Jobs on the iPhone. Jobs had no idea that the search giant was secretly working on its own smartphone. The first Android smartphones were launched a year after the iPhone and by 2011 were outselling the far more expensive Apple product. In 2021 there are about 3.8 billion smartphones – Apple and Android – in use around the world[3] serving almost half of the world's population. In the most advanced markets well over 90 per cent of people have smartphones.[4]

Everywhere, Present Day

Look around wherever you are and you'll see people staring intently at the small screen in their hand. In just over a decade we have become addicted to these magical windows on the world. Today, smartphones are much more than telephones, iPods and internet browsers. They are also clock, diary, torch, scanner, camera, world atlas, road map, calculator, compass, pedometer, credit card, accounts book, tape recorder, tape measure, tracking device, TV, radio, computer, notepad, library, photo album, spirit level and games console. As you read this the list will be growing.

Smartphones have allowed and, in some cases, forced more and more aspects of our lives to take place on screen. Shopping, banking, arranging travel, booking appointments, moving house and applying for jobs have all become activities conducted on screen rather than in person or over the telephone. Meanwhile, there are huge companies and whole industries providing products and services that only exist because of smartphones.

All this makes it ever more necessary for us to keep our smartphone to hand at all times – and should we dare to ignore it for a while it will soon demand our attention back with lights, vibrations, pings and beeps.

British communications regulator Ofcom reports that young Britons spend at least as much time online on their smartphones and tablets today as their parents spent watching television in previous decades, and time spent online now exceeds time spent watching television for adults overall. For many young people, checking their phone is the first thing they do when they wake up and the last thing they do before going to sleep, while one in five wakes up regularly to check in overnight.[5] The measures taken to combat COVID-19 in 2020–21 have made millions of us into self-declared smartphone 'addicts'.[6]

Steve Jobs described phone calls as the 'killer app' of mobile devices. This is no longer the case. Time spent on mobile phone calls fell in 2017 for the first time, and most people now say that looking at their screens is more important than talking on the phone.[7]

Anytime, and almost anywhere, smartphone users can see the pinnacle of culture, the depths of human depravity, and everything in between. We can browse the world's greatest museums and galleries and peruse something approaching the sum of human knowledge without taking a step, entertain ourselves for hours playing games, watching films and TV programmes and scrolling through endless streams of cat videos, fill our minds with advice and opinions from pretty much anywhere, and follow news events as they unfold.

In among all the positive content are things we don't want to see, and certainly don't want our children to see: graphic violence, hate speech, pornography, and sites promoting self-harm and suicide. Filters are not always able to block these out, prompting calls for legislation to put the onus on the companies that host such content to remove them promptly. The biggest companies – Google and Facebook and their subsidiaries including YouTube and Instagram, with others, collectively called Big Tech – have so far managed to argue successfully that they are merely platforms rather than publishers, and as such not editorially responsible for the content on their sites. With pressure from governments and consumers mounting, especially after widely reported personal tragedies, this may well change in the near future.

A Social Society?

Hand in hand with the phenomenal take up of smartphones is the worldwide explosion of social media. The concept of informal online networks originated in the 1970s with chatrooms and bulletin boards. In 1997 a website called SixDegrees combined user profiles with lists of connections or friends that other friends could see,[8] and others followed. Some endured, including business-oriented site LinkedIn, founded in 2002 and now owned by Microsoft. Others burned out, like Friendster. The early social networks MySpace, Bebo and Friends Reunited were sold for hundreds of millions of dollars in the early 2000s, making what seemed like fortunes for their founders, though the sums are chicken feed compared with the billions made by later internet entrepreneurs. Only MySpace survives today, as a shadow of its former self.[9]

Facebook began life in 2004 as a Harvard University student directory run from Mark Zuckerberg's dorm room. By 2006 it had opened its membership to anyone and was growing fast. The company launched a mobile-optimised website the day after Steve Jobs presented the iPhone at Macworld. Facebook user numbers and smartphone ownership rose together, each stimulating the other. By 2008 Facebook had 100 million users; four years later it topped a billion. There are now more than 2.8 billion Facebook users active every month,[10] and another billion on its sister site Instagram, the overwhelming majority of whom access the sites from smartphones.[11] Other major social media sites around the world, also accessed from smartphones, have billions more active users.[12]

San Francisco, 2010

On the same San Francisco stage, dressed in the same outfit and again attended by thousands of faithful geeks, Steve Jobs launched the iPhone 4 in June 2010. Its show-stopping innovation was a video-calling function he called FaceTime. Jobs FaceTimed his buddy Jony Ives from the stage, and the two chatted breathlessly about how they'd dreamed of video phones since watching *The Jetsons* on TV in the 1960s. They cooed over the new phone's front-facing camera, added to the usual rear-facing camera to allow users to see each other on screen while making FaceTime calls.

The new camera proved useful for more than FaceTime. The front-facing camera allowed people to see themselves on screen *and* take a photograph at the same time. Almost immediately people started posting self-portraits on social media, and a new phenomenon was born. Just like the crazes for daguerreotype portraits and *cartes de visite* more than a century and half earlier, the world went mad for smartphone selfies. Everyone wanted to see themselves and, it seems, for others to see them, too. Selfies were taken in space, in the midst of wars and natural disasters, with popes and presidents, by monkeys and movies stars. Dozens of people died trying to take 'ultimate' selfies.[13] In 2013 the Oxford Dictionary named 'selfie' its word of the year.

Social media was and remains the engine of the selfie craze: the posting is the point. More than 90 per cent of American teens were posting selfies online in 2013, and the trend hasn't abated. Today the users of selfie-based message app Snapchat send over 5 *billion* snaps a day.

So, what is the perfect selfie? The British fashion photographer Rankin took photographs of fifteen teenage girls and asked them to make them more 'social media friendly'. The girls – all of whom could easily be described as beautiful and were shot in flattering light by a famous photographer – typically gave themselves larger, almost alien eyes, smaller noses, fuller lips and hyper-luminescent skin. Rankin was shocked by the result, despite being famous for Photoshopping celebrity images himself. But this type of hyper-idealised editing has become entirely normal. There are multiple apps available that make it possible for anyone to achieve Photoshop-style touch-ups previously available only to professionals, on their phones. The trouble is, as Rankin found, these tools are often taken to extremes, used to distort rather than to polish, and create images that real life can never match.[14]

Several commentators have claimed that the selfie craze has caused an epidemic of narcissism in twenty-first-century society. One calls it the 'other' epidemic of the modern age, the first being obesity.[15] While other scholars have challenged this analysis,[16] it is difficult to argue that the endless carefully edited and touched-up photographs of sculpted abs, perky rears and perfectly made up pouty faces that feature on thousands of social media accounts are not evidence of narcissistic tendencies on the part of their subjects. But narcissism is a broad term. It spans normal and healthy human self-regard, annoying expressions of vanity, and a serious personality disorder.

While social media may provide a platform for narcissists, psychologists have observed narcissism among young people (as measured by the Narcissistic Personality Inventory*) rising steadily for about thirty years. No one knows exactly what is behind this increase, but it predates social media by decades. Nevertheless, researchers who studied a group of university

---

* The Narcissism Personality Inventory is a series of questions widely accepted among researchers to determine relative levels of narcissism. You can find out how narcissistic you are by taking the Inventory yourself at openpsychometrics.org/tests/NPI.

students over a four-month period in 2018 did find that those with signs of problematic internet use* who also used visual social media platforms such as Facebook and Instagram showed an increase in narcissism over the observed period. The researchers couldn't say definitively what caused the increase but, perhaps significantly, students with similar internet usage but who only used text-led social media sites such as Twitter *didn't* show an increase in narcissism over the period.[17]

Offline (or IRL,** as young people might say) there is evidence that the obsession with visual media has entered our personal space. Both men and women from all walks of life are spending more time and money than ever on their appearance. According to one alarming estimate, achieving the perfect eyebrow costs British women £2.3 billion a year, and that's a tiny proportion of the staggering £300 billion women spend on grooming overall.[18] Men are not immune to appearance-based pressure either; male grooming today goes way beyond a shave and a haircut.[19] Young men routinely wax, pluck, moisturise, and cover up blemishes their fathers might never have noticed. Concealer can't be the only product they buy: the market for male grooming products is worth nearly $60 billion worldwide.[20]

For those wanting more than a temporary fix, extreme measures are available via the booming cosmetic surgery industry. Nearly 25 million cosmetic procedures were performed worldwide by plastic surgeons in 2019, half as many again as in 2010, and undocumented millions more had botox and fillers from less qualified practitioners. Breast augmentation is the most popular surgical procedure, while the fastest growing

---

* PIU, as defined by criteria such as increased tolerance over time, obsessive behaviour, withdrawal symptoms and inability to control use.
** In real life.

is vaginal rejuvenation, up 23 per cent in 2017 compared with 2016.[21] The increase in awareness of 'sexual aesthetics' is fed by the normalisation of pornography accessed online and pressure among young people to share 'sexts' – a subset of the selfie and another unforeseen use of the smartphone camera.

Younger people are going beyond hair and facial embellishment and are decorating their bodies, too. Tattoos and body piercings are practically compulsory for the under-30s.[22] All this primping and preening is skilfully posed, captured, edited and posted in the millions of selfies that go online every day.

Everywhere, Present Day

Smartphones have enslaved our eyes like nothing before them. There are all the practical uses we put them to in our daily lives – shopping, banking, and looking up information – but the sites that tend to draw people back again and again are the social media platforms. They meet a deep human need to connect, to be seen and acknowledged. These are primal needs, ingrained in our psyches ever since we gathered around campfires. Arguably these needs had been unmet in recent decades, ever since everyone went inside to watch TV, or perhaps the lack of human connection goes back even earlier, to the time when we moved away from small rural communities into large, anonymous cities. Either way, it is clear from its present ubiquity that social media appeals to people in a deeply visceral way.

Not only is social media inherently attractive, but Facebook and other social media sites have spent the last decade carefully tweaking their offerings to make them irresistible. The first major innovation was the introduction of the News Feed, a running list of every bit of new information about a user's friends, constantly refreshed. It removed the need to look up friends one by one and made it more enticing to check in with the site, and doubled the time users spent on Facebook.[23] It also allowed Facebook to insert ads directly into the flow of a user's feed rather than have them sit to one side. Allowing users to add comments and replies into the News Feed gave people another reason to check in.

Another game changer for Facebook was the 'Like' button, introduced in February 2009. Likes gave users a very low-effort method of acknowledging and interacting with friends' content and dramatically increased the number of reactions people got to their posts. This proved highly addictive, as it exploited both the psychological need for connection and social affirmation *and* physical brain chemistry. As Sean Parker and other former Facebook executives have since explained, likes and comments are little rewards that trigger a release of dopamine in the brain. This is a feel-good chemical that is also released when we eat nice food, have sex or take cocaine. It can encourage positive behaviours but it is also closely associated with addiction. Every time a social media user gets a like for their post, they get a little dopamine hit. Brains are quick learners, and they like dopamine hits, so users are encouraged to post ever more eye-catching images, and to check their smartphones more and more often, hoping for another hit.

Likes also allow Facebook to tweak what its users see, so that the things they are most interested in appear first. The more likes a user gets, the higher their posts appear in their friends' News Feeds, earning them even more likes. Posting, liking, looking, getting liked, commenting, checking in, etc., is a cycle of reward-seeking called a dopamine feedback loop, and Facebook and others have devised all sorts of clever ways to bring users back into the loop when they stray. Notifications –

the pings, buzzes and lights that indicate something you need to see – are one method. They trigger a little dopamine hit as we wonder what that little buzz means. That is why apps send you notifications as often as they can find an excuse for it – even spurious excuses such as 'friendiversaries', unless you get around to turning them off. These are all carefully doled out to maximise their neurological impact, adopting the same 'variable reward' strategies casinos use to keep players feeding coins into slot machines. Researchers have found that our brains find unpredictable rewards particularly alluring, so Facebook will often withhold your likes for a while before delivering them in a bundle, to boost the dopamine effect.[24]

Young Lives Online

Sociologists have agreed for more than a century that young people define themselves to an extraordinary degree in relation to their peer groups, and in particular by how they are seen by others.[25] The adolescent view of their own 'self' is the 'self' that others see.[26] Thus the mechanisms social media provide for social connection and approval are especially attractive to adolescents and young adults. It is no surprise that social media began with young people, or that this is where its centre of gravity remains.

This is also a potential problem for young people and society, however. A vital part of the transition from youth to adulthood, sociologists tell us, is stepping away from defining oneself in terms of others and becoming the independent author of one's own life.[27] The concern among researchers is that social media traps young people in the web of seeing themselves purely as others see them, without the tools to develop a stable sense of self, defined in one's own terms.

An even more prevalent concern is that because of the natural tendency of young people to define themselves through the eyes of others, they are particularly prone to comparing themselves with others. Social media provides previously unimagined opportunities for social comparison; indeed once the like button was introduced, comparison became the very essence of the social media experience.

It was well established long before smartphones that social comparison can make people unhappy. Richard Layard's classic 2005 book, *Happiness*, is a highly influential study of the sources and nature of happiness in societies. Early in the book, Layard cites a famous Harvard experiment in which students were asked to choose between two imaginary scenarios. In the first scenario they received $50,000 while others got $25,000. In the second they received $100,000 while others got $250,000. $1 bought the same amount in both scenarios. The majority of students preferred the first option. That is, they chose to get less money overall if that meant they received more, rather than less, than their peers.[28] *

The point of the experiment was to illustrate a curious aspect of human nature. Whatever level we are at – and this operates financially, materially, physically and emotionally – we are reasonably happy if those around us are in the same boat. We are even happier if we are slightly better off than our neighbours. But once it becomes clear that others are outdoing us, happiness goes into decline. Gore Vidal put it succinctly when he said, 'every time a friend succeeds, a little piece of me dies'.

Consider this in relation to media. Before television, social comparison was limited to what people saw in their own

---

* The subjects were students at the School of Public Health. I wonder if the outcome would have been the same had they been at the Harvard Business School?

neighbourhoods. When television came along, people started seeing idealised versions of people and their lives in TV shows and advertisements that looked better than their own. There was a rise in theft at first as people saw what others had that they didn't.[29] But over time people got used to it, knowing that the worlds of the cinema and TV are mostly fictional, even – or perhaps especially – in so-called reality TV.

Enter social media. Now people see what purport to be real-life images of the perfect lives of people they know! Not only do they look great, they are having more fun, with more friends, nicer clothes, better abs/butt/lips/hair, more expensive holidays, and so on. Even when people know that they filter, tweak and embellish their own profile and posts, it is very difficult to apply an equivalent discount when looking at others'. The impact is bound to be negative, especially for young people, for whom social comparison has a disproportionate importance.

There has been a flurry of academic studies on the impact of social media on young people's mental health, and the news isn't good. Study after study has linked social media to declines in mood, increased body consciousness, compromised well-being and generalised declines in mental health.[30] University of San Diego psychologist Jean Twenge cites a wide range of statistics to support her claim that the increase in time young people spend looking at smartphones and social media has caused a dramatic increase in teenage unhappiness, mental health problems and suicides.[31]

Not everyone agrees with these arguments,[32] and when asked about smartphones, young people themselves seem to speak up for the device accused of sending them spare. A 2018 survey by the respected Pew Research Center found that a majority of teens believe social media has had a positive impact on their lives, saying it makes them feel more connected to their friends' lives and more in touch with their friends' feelings. At the same time, almost half complained about all the 'drama' on social media and 40 per cent said they feel pressure to only post content that makes them look good, or that will get lots of comments or likes. Only about a quarter said social media has had a negative impact on their self-esteem, and just 4 per cent said it makes them feel 'a lot' worse about their lives.[33] Of course, modern teens who have grown up with social media have no way to assess how well it connects them to their friends compared with other old-school activities like, say, talking on the phone or hanging out at the mall.

Whatever the underlying cause, statistics do suggest a rise in mental health problems since the advent of smartphones and social media, especially among the young. In the UK, NHS data reveals increases in the numbers of people seeking help for conditions such as depression and anxiety,[34] while ONS figures show a 45 per cent increase in suicide among young women and girls, 10–29 years old, between 2010 and 2017.[35] A *British Medical Journal* study published in 2017 said self-harm among girls aged 13–16 had increased by two-thirds between 2011 and 2014.[36]

Similarly, a US report published in *JAMA*, the *Journal of the American Medical Association*, found no significant change in levels of self-harm from 2001 to 2008, but a trebling of self-harm among young girls between 2009 and 2015.[37] Another US study found an increase in rates of major depressive episodes among 12–20-year-olds between 2005 and 2014,[38] while yet another showed a 24 per cent increase in the youth suicide rate between 1999 and 2014, with the increase greater after 2006, especially among girls aged 10–14.[39]

While these statistics don't pinpoint a cause, they certainly suggest something has been making young people unhappier and less stable over the last decade.

An Antisocial Society?

Where people sat passively watching TV for hours, the smartphone offers, and demands, much more. Where eyeballs in the room were enough for television advertisers and executives, internet content providers and advertisers demand *engagement*. In the online space, simply seeing an image is just the beginning. For it to count in commercial terms it needs to be liked, favourited, shared, clicked through, linked to, subscribed to, commented upon, tagged, or otherwise interacted with. If an advertisement on Facebook doesn't achieve this it is quickly dropped, for fear of interrupting users' ever more tailored, ever more 'enjoyable' (read addictive) experience.

Where broadcast television must appeal to the widest possible audience, and therefore tends towards the somewhat bland path of least resistance programming, online content is the opposite. It seeks to inflame users – positively or negatively, it doesn't matter – in the name of user engagement. Where television executives valued inactive, passive audiences who kept watching, online content providers are most rewarded by material that stimulates their audiences into action. This has prompted widespread concerns that social media algorithms are leading to polarisation and division in society, as the most extreme views end up being the ones most widely distributed.

Customer or Product?

The price we paid to watch free broadcast television was to watch advertisements. Social media also demands that we watch ads, but it is watching us, too. Around the time of its IPO in 2012, Facebook started experimenting with the reams of data its users generated as they used the platform, under the guidance of its new Chief Operating Officer, Sheryl Sandberg. Sandberg had come from Google, who were already experts at collecting and analysing user data and finding new ways to exploit what Harvard professor Shoshana Zuboff calls 'behavioural surplus'.[40] This is all the extra information that digital companies collect in the course of providing a basic service – things like time of day, location and abandoned purchases. Facebook was literally teeming with potential behavioural surplus. They knew when users interacted most, with whom, what they liked, what caused the most reactions, what sort of language they used, their political views, and a host of other personal details via their social media profiles and interactions. Much of this information had been ignored or discarded up to this point, but Sandberg realised it was Facebook's most valuable asset. The company began discreetly collecting and analysing user data in new ways and combining it with even more data acquired from other sources via secretive data-brokers.

Today, data collection and sharing goes on across all our digital activities. This extends from online search and social media to all those useful smartphone functions that replace practical, physical tools: these apparently innocuous apps happily send your and my user information to Big Tech. Social media users are not only the customers of the platform providers, they are also the product.[41]

The Big Tech companies now routinely track users' every move to gather reams of data about their daily habits, attitudes and activities. This ranges from relatively banal, though still private, information like what people do when, where they shop, and when they go to bed, to highly personal details about political and religious affiliations, relationships and health. Hardly any of this gathering is apparent to social media users, although they have probably given permission for their lives to be harvested via the lengthy user

agreements they consented to but never read. User information is combined and processed by proprietary algorithms that
allow advertisers to target potential customers with great preci-
sion, and to tailor their messages to an individual's preferences
and prejudices to maximise their influence. The aim is to know
you better than you know yourself. There is already anecdotal
evidence that this is entirely possible,[42] and many of us have
experienced mysterious online coincidences that suggest our
devices might know more than we think.

The trouble was, or rather is, that it's not just consumer goods
companies using these tools to get us to buy a few more products. Political parties and interest groups on both the right and
left have used social media algorithms to influence recent elections in the US and UK, well below the radar of conventional
– and regulated – political advertising. The Kremlin-associated
Internet Research Agency used social media tools to influence
the same elections in the West, and has more recently inflamed
tensions in the US and elsewhere purely for the purpose of
troublemaking and disrupting the idea of democracy.

It doesn't require the resources of a government, a cor-
porate ad budget, or even a lobby group to exploit these
algorithms. Disinformation, fake news and hate speech have
been successfully spread online for modest commercial motives
perfectly indifferent to politics or ideology. An obscure town
in Macedonia called Veles set up a cottage industry producing
fake news in the run-up to the 2016 US election, producing
ever more salacious stories in the hope of automatically earning advertising revenue via the Google AdSense algorithm.
The most successful Veles fake news entrepreneur, aka Boris,
set up several pro-Trump Facebook groups and linked them
to his websites, where he published daily pro-Trump stories
hacked from alt-right sites in America. He used dozens of fake
Facebook profiles to like and share his posts, promoting them

up the system via Facebook's algorithms. By diligently tending his crop of websites and Facebook groups, Boris earned thousands of dollars over a few months, until he was eventually shut down by Google just before the election.[43] It wasn't a fortune, but what an unemployed young man did in pursuit of an easy income may have influenced the election of the leader of the Free World.

Professor Zuboff describes the business models of the Big Tech companies like Google and Facebook as 'surveillance capitalism' and documents the uneven exchange between users and providers in great detail in *The Age of Surveillance Capitalism*.[44] Zuboff asserts that whereas industrial capitalism involved the exploitation of nature for profit, surveillance capitalism is the similar exploitation of *human* nature. She calls on governments and citizens to fight back against the new robber barons of Silicon Valley who quietly mine our private lives and thoughts for their own purposes, even while their own activities remain largely invisible to consumers and regulators.

What are we to make of all this? Are we in fact all lab rats willingly participating in a giant global experiment, being observed through a discreet one-way mirror of which we are only just becoming aware? As we look anxiously at our screens, again, desperately hoping to find a few likes or a friendly comment among the torrent of carefully targeted ads, our actions are being tracked (perhaps down to the very movements of our approval-hungry eyes), analysed and fed into the data machines that will make sure our next visit will come even sooner, and be even more irresistible.

The tide seems to be turning, with major government studies and court cases under way that may result in curbs to Big Tech's power. The European Union introduced the General Data Protection Regulations in July 2018, and the UK Government has published a White Paper and Response on Internet Safety suggesting various measures including appointing Ofcom as 'Online Harms Regulator'. Change may be a struggle, however, in the face of a society in the grip of a collective case of Stockholm Syndrome. Despite the spate of reports and articles expressing concerns about Big Tech capturing, selling or even losing personal information to those who want to manipulate us for profit or more sinister purposes, and scandals such as Cambridge Analytica, enthusiasm for social media has waned … hardly at all. And for those who have tried to leave Facebook, some have found it so tightly intertwined with important aspects of their lives that this was easier said than done.[45] The more Facebook and the like insinuate themselves into our lives the harder it becomes to extricate ourselves from them. For many people, Facebook and friends are already too big to fail.

AN ALL-SEEING WORLD

Nothing can cure the soul but the senses,
and nothing can cure the senses but the soul.

Oscar Wilde (1854–1900), *The Portrait of Dorian Gray*, 1890

Screens have made the modern world more visual than ever before. From the moment we open our eyes in the morning we are looking, looking, looking, because there is always something to look at. It's not just smartphones. Electronic screens are ubiquitous, in homes and offices and also in buses, taxis, elevators, on shopping trolleys, billboards, lavatory doors and walls and even underfoot. They compete endlessly for our eyes' attention.

Off screen, the jostling for visual attention continues. Advertising is squeezed onto every available surface, and shops offer their wares in ever-changing window displays. Companies spend fortunes on branding and packaging to coax our eyes towards their products. Even fruit and vegetables have to meet strict standards of visual perfection, while modern packaging makes it impossible to smell, squeeze and taste fresh food as our grandparents did.

Seeing and being seen dominate twenty-first-century life. Meanwhile, our other senses are gradually being sidelined, relegated to the off-duty domains of pleasure and leisure. It is worth considering what impact this concentration of our five senses into one – the ever narrower funnelling of human experience through the single channel of vision – is having on our lives …

Hearing was once our lifeline. Ears can hear from any direction and over long distances, and were much better than eyes at detecting danger or enemies. Sound was also our principal method of communication for hundreds of thousands of years, right up to the spread of printing and literacy from the fifteenth century. Since the noisy incursion of machines into our lives during the Industrial Revolution 200-odd years ago, we've trained ourselves to block out as much sound as possible. Today, listening for entertainment remains popular but our ears are called on less and less for practical purposes throughout the day. People spend their time inside sound-proofed buildings, where the natural sounds of birds, weather, and passers-by are inaudible and irrelevant. Sounds are, more often than not, seen as noise: a nuisance.

Person-to-person communication was, for most of human history, a multisensory activity, and primarily aural. Today, direct communication is more and more often circumvented by technology. There's no need to ask for directions when there's a digital map in your pocket. Nor is there a need to talk to a fellow passenger once a screen tells you when to get off the bus. You can opt for self-checkout at the store, or 'better' still, shop online and avoid direct human contact altogether.

When we do communicate directly with other people, many of us now find it easier to do so silently. We can share our news or discuss the issues of the day by mutely posting words and pictures on social media. We can respond to our friends' posts without even having to find words – a like or an emoji does just as good a job as finding the words for a comment.

Even one-to-one communication is going quiet, moving from voice to noiseless text. Time spent on phone calls is falling overall. British 16- to 24-year-olds spend ten times as long communicating via text-based services as they do speaking on the telephone,[1] and the same thing is happening all over the world.[2] Let's face it, who hasn't sent a text message instead of talking to someone face to face, even when they're in the next room?

But text is not equivalent to talk. Speech and vocal signalling are ancient and primal human activities. Written language is only a few thousand years old. Writing is not hardwired into our brains like spoken language, so we don't just pick it up as babies, like we do with speech. We have to learn reading and writing and they remain higher order human skills.

In a 2011 experiment, a group of girls aged between 7 and 12 were each subjected to a stressful situation – they had to give a presentation to strangers. Afterwards, a quarter of the girls were allowed to see their mothers face to face, another quarter to speak to their mothers on the phone, the third to text back and forth with their mothers, and the fourth group to sit quietly alone. The researchers measured the girls' stress hormones over time. The girls who talked to their mothers on the phone calmed down as much as the girls who interacted with their mothers directly. The girls who texted with their mothers calmed down as *little* as the girls who sat alone. As good scientists, the researchers didn't presume to extend their conclusions beyond their sample group, but the subtitle of their paper gives their core conclusion away: '… why we still need to hear each other'.*[3]

---

* Incidentally, of the many dozens of studies I read researching this book, this is the one that has most significantly changed the way I feel and behave.

Making rhythmic noise together was a vital aspect of community bonding and religious practice for millennia. Chanting fans at a football match carry on this tradition and get the feel-good endorphin rush associated with it, but the opportunities for such group activity were becoming scarcer even before COVID-19. These are such powerful tools for group empowerment that marching, chanting and singing are often frowned upon, and even banned in some circumstances, even in their most basic expressions. The British National Union of Students discourages hand clapping, whooping and cheering at some of its events, preferring the visual expression of approval via silent 'jazz hands' as used in British Sign Language. Some junior schools use 'marshmallow claps' – hands not touching – when celebrating success. These measures are intended to avoid upsetting people with certain disabilities. Valid aims, but one of life's visceral pleasures denied.

It is well known that, along with the voices of loved ones, natural sounds such as wind, rain, running water and waves are deeply calming. For those who can't access the real thing, the internet is crammed with recorded sounds of nature, packaged to soothe the soul. Music – another primal activity that was once a practical tool to encourage bonding and aid memory but is now relegated to the sphere of entertainment – can also evoke a sense of contentment. In a somewhat ironic development, social media has invented or discovered a new type of sound that can also be deeply relaxing. ASMR stands for Autonomous Sensory Meridian Response and refers to a tingling sensation in the head and back that some call a 'brain orgasm', brought on by certain sounds. There are something like 13 *million* ASMR videos on YouTube, some of which have had millions of views. Most feature a whispery voice followed by abstract but real sounds like a gently tapping fingernail, light scratching, a cup being pushed along a table, or paper being crinkled. It doesn't do anything for me but my children tell me some of their friends find them extremely helpful for calming down when they are stressing out at school or for getting to sleep. Whether they would do better to open their window and turn off their electronic devices is a moot point.

Smell was another vital alarm system for our ancestors. It directed them towards potential food sources and mates, and away from enemies and danger. Humans have highly developed olfactory capabilities – in some circumstances we can out-smell dogs.[4] What's more, the area of our brains responsible for smell links directly to the parts of the brain responsible for memory and emotions – none of our other senses has the same direct connection.[5] Despite this, smell as a vital attribute is largely ignored by modern life. Polite society banishes all but the most banal scents from our experience, as we deodorise ourselves and our environments, and employ efficient fans to remove the smells we can't avoid. The scents we do allow ourselves – soaps, perfumes, candles, room sprays, fabric softeners and the like – are artificial and redundant, the equivalent of white noise for our noses.

Food smells have reduced significantly over the past few decades as products have been ever more heavily packaged, sealed and refrigerated. At the other extreme, specialist scent marketeers are infusing environments with artificial odours designed to manipulate consumers. A 2011 *Time* magazine article caused a stink when it revealed that a grocery store in Brooklyn was pumping the air full of artificial chocolate and bread smells in an effort to entice customers to stay longer and buy more.[6] They are by no means unique. M&M World stores deploy fake chocolate aromas because all the real M&Ms are too tightly sealed in packages to emit any odour. A British packaging company called Scentmaster promises to supply plastics containing what its website calls 'addictive technology' that can 'mean the difference between the stench of failure or the sweet smell of success'.[7]

A 2016 study found a very strong link between loss of the sense of smell – called anomia – and depression. Patients with depression tended to have reduced smelling abilities and, similarly, patients with reduced smelling abilities were more likely to be depressed. Depressive symptoms worsened with the severity of the loss of smell.[8] This patently strong link between smell and depression raises at the least the possibility that the modern lack of access to the natural, primal smells our ancestors took for granted – soil, grass, rain, sea, animals, our own bodies, fresh food, each other – could be contributing to our declining dispositions. In an anecdotal example of smell lifting mood, I cite my youngest daughter, who thrusts her face into our dog's ear and inhales deeply whenever she feels sad or upset.

Taste is also pretty much a liability today. Our sense of taste evolved to recognise the most nutritious foods. For our hungry ancestors, this meant those with the most calories. That is why we love sweet, starchy and fatty flavours, and the salt that amplifies them, but this is hopelessly outdated in a world of plentiful, industrially produced food. Food producers know this and lace their products with the sugar, fat and salt our prehistoric taste-buds find almost impossible to resist, producing an international obesity epidemic. The *New York Times* food editor describes a major recent food trend, 'Dude Food', as 'anything that is salty, fatty and crisp, with bro-tastic sweetness and wicked heat'.[9] Yum! Incidentally, our modern love of spicy heat may be the last bastion of useful taste sensations. Spices have powerful anti-bacterial properties, which is why warm countries tend to have spicy cuisines.[10]

People who manage to break the stodge/salt/sweet habit are rewarded with more than just slimmer bodies. An Australian study demonstrated that increasing fruit and vegetable consumption led to significant increases in happiness, life satisfaction and well-being. They found that eating eight extra portions of fruit and vegetables a day increased life satisfaction as much as going from being unemployed into employment.[11] That's a massive happiness boost – surely a good enough reason to try to override our prehistoric desire for sweet, salt and fat.

With the exception of tapping and swiping screens, our fifth sense, touch, is suffering no less. All sorts of tactile objects such as keys, taps and switches are being replaced by screens. Interaction with many of the varied shapes and surfaces we used to encounter in a day's activities are being superseded by technology. When was the last time you wound down a car window or pricked yourself with a needle while sewing? Will you turn this page when you reach the bottom, or swipe a screen?

Touch screens are big news in education, but there is ongoing debate among educators about their relative benefits and risks. Handwriting is a particular hot topic. Finland recently announced that it is dropping cursive handwriting lessons for children, and in the US, the Common Core Standards no longer require schools to teach cursive script, although many continue to do so. Meanwhile in Australia, occupational therapists report that more and more children are starting school unable to hold a pencil.[12] Educators in several countries have expressed concern that some children are not learning fine motor skills in their early years because they spend their time playing on screens rather than with real toys and crayons.[13]

The devaluation of physical feeling extends to our whole bodies. Modern clothes, infused with materials that make them stretchy and soft, are the tactile equivalent of invisible. They regulate heat and cold and accommodate our ever-increasing girths. Our skin is almost never troubled with the squeezes, scratches, stiffness and strappings of our ancestors' garments.

Clad in our comfy, stretchy clothes, it is perfectly possible to pass an entire, productive day without moving much at all. Air-conditioned vehicles traversing smooth roads have replaced walking and riding from place to place. Natural sensations such as wind in the hair, rain on the face, a sudden drop in temperature or sun in the eyes are the stuff of holidays and historical novels, banished from everyday life as inconveniences.

Everyone knows that people like and need to be touched. As if it needs proof, a 2016 research report found that receiving affectionate touch promotes relational, psychological and physical well-being in adults. The researchers found that touch not only reduces stress, but also promotes well-being independent of stress.[14]

Despite this general and academic knowledge, we are having less physical contact with each other. Person-to-person touching is ever more proscribed in the West, culturally and, increasingly, legally. This is a response to appalling past incidents, but the result can be a deficit in what most would recognise as normal human contact. Overzealous implementation of 'don't touch' policies can be seriously detrimental. Old people in care homes and people living alone can end up having no physical contact with others at all. Psychologists have even come up with a name to describe the paucity of human touch that many people now experience: 'skin hunger'. Was a more desperate phrase ever coined?[15]

This concentration of our five senses into the almost exclusively visual extends to the most visceral of human activities. Think of war and think of sex. Participation in both of these can now be, and often is, completely removed from the physical, sensate realm and conducted through sterile, obedient screens. The images are graphic, sound is optional. Smell, touch, and taste are nowhere to be found. Killing and loving become equally bland activities.

Holistic Happiness

I am acutely aware that this chapter is in danger of turning into a hippy's charter. Go outside! Smell the roses! Hug your neighbour! Talk to your children! Nevertheless, I offer this analysis in all seriousness. Our bodies and brains evolved to interact with the natural environment, and each other, through all our senses. I believe it is possible that the modern world's relentless channelling of the functions of day-to-day life through our eyes alone is causing some of the desperation and disconnection people feel. As we've seen, there are strong positive associations between each of our non-visual senses and happiness. The dramatic imbalance in our sensory experiences could be making us unhappier. It is surely worth exploring that possibility further.

Peak Seeing?

Of course, there are still plenty of people who make extensive and beneficial use of highly developed non-visual senses. And millions of blind and visually impaired people lead happy, active lives as fully contributing members of society with very little or no vision at all.

Nevertheless, it remains the case that, especially among younger, urban populations in the developed world, the present dominance of visual stimulation in the smartphone/social media landscape goes beyond anything seen before at any time in history. The world we can see though our smartphones presents huge opportunities, but also substantial risks. And then there's the question of who is looking at us.

How Did We Get Here?

The history described in this book tells us that the ascendancy of seeing began long, long before smartphones and long before television or electricity. The seeds of the twenty-first century's relentlessly visual culture were sown a million years ago when our proto-human ancestors tamed fire and learned to see in the dark. When people started making images on walls and carving figures from pieces of wood and stone 40,000 years ago they established the beginnings of a culture that would eventually be dominated by the visual. The Mesopotamian scribes who invented writing 5,000 years ago took depiction a dramatic step further and precipitated the decline and eventual fall of oral traditions. Other man-made visual technologies fundamentally changed what and how people saw themselves and their worlds. Each one in turn had profound effects on the human psyche and society, and changed history in ways unforeseen at the time.

Have we reached peak seeing? Perhaps we should have. The blue light that our indispensable electronic screens emit can harm the treasured eyes that spend so much time looking at them. Ophthalmologists are concerned about links between excessive blue light exposure and increased or earlier incidence of macular dystrophy and cataracts, both of which can lead to reduced sight and, occasionally, blindness.* Blue light also

---

* Apps such as f.lux can filter blue light from computer screens, while some phones now have blue light filters available in their settings.

suppresses the production of melatonin, which can mess up sleep patterns and has been linked to depression, Alzheimer's, obesity and some cancers.[16]

This is not a case against digital technology, social media or modern culture. There are countless ways individuals, institutions, and society in general are better off today than ever before. In any case, the genie is out of the bottle. I'm no more immune than anyone else to the time-wasting temptations placed before me, and equally addicted to my own screens – small, medium and large. Apart from anything else, I could never have undertaken the project of researching and writing this book without them.

My aim has been to tell the story of how I believe we got to where we are. Humankind is extremely resourceful and also very impatient. From the very beginning, the eyes that nature gave us were not enough and, again and again, we sought to see more than our feeble bodies allowed. With each visual invention and innovation mankind came up with the world, seen differently, became a different world. And with each invention, unintentionally but ultimately, seeing became a little bit more important.

The smartphone story is the latest contributor to a story that has played out many times in the long history of seeing. It is because of the immense impact that changing what and how we see has had time and again throughout human history that I believe we need to approach our future relationship with visual technologies with some caution, or at least with our eyes open. For a start, we need to take care of those precious, precious eyes. Switch off our screens and go outside. No artificial light beats natural daylight's mood and health-boosting properties. Just looking away from the screen for a few minutes gives our eyes a break from the blue light, and eyes are more relaxed when looking at distant objects.

We need to attend to our other senses, too, and do more talking, touching, tasting and smelling every day. They are still in full working order, after all, even if they have been a bit neglected lately. I believe that shifting focus away from the eyes will bring immense pleasure and richness to our lives, and may well make us happier and healthier.

Even more importantly, we need to fully embrace embodied relationships with each other, by which I mean taking on the human experience we are made for: talking to and moving with and touching and even smelling other people, just for the joy of being alive.

Finally, I hope we can find more time to embrace seeing in a new – old – way: staring into space, looking at nothing in particular, letting our minds wander and our thoughts take us where they will. Who knows what we might see there?

EPILOGUE:
2019

The February 2015 phenomenon of *#thedress* wasn't just a brief social media sensation; it has continued to fascinate vision scientists around the world and has prompted dozens of academic studies and papers. It turns out the image was extremely rare, in that it happened to have exactly the right absence of information to create significant visual uncertainty.[1] That is because when our brains interpret colour they assess the signals they receive in comparison to surrounding colours, and subconsciously factor in things like background light levels, and shadows and highlights, based on a lifetime's experience and memory. This is called colour constancy and is a critical feature of visual perception. In the case of *#thedress*, there were no surrounding colours and very few visual clues to help viewers interpret light sources, forcing viewers' brains to make unconscious assumptions.

A study published in the Journal of Vision in 2017[2] surveyed over 13,000 participants, asking them a series of questions about their perceptions of *#thedress*, as well as demographic and lifestyle questions such as whether they spend a lot of time outdoors and whether they would call themselves a morning- or evening-oriented person. The survey confirmed empirically an hypothesis posed by several other researchers. In the absence of any other visual clues, some people automatically assumed the dress was in the shadow of a strong, blueish light, such as daylight, and so adjusted out the picture's bluish shades to interpret the dress as white and gold. Others assumed the dress was in warmer, artificial light and automatically adjusted out the yellowish shades in the photograph, leaving the dress they 'saw' blue and black.

What's more, the study showed that lifestyle had a significant impact on these unconscious assumptions, and therefore how people interpreted the image. Morning-loving larks, who presumably are more typically exposed to daylight, tended to assume the image was taken outdoors in shaded sunlight. The dress thus became, in their eyes, white and gold. Night owls, on the other hand, tended to see a dress lit by lightbulbs that was, to them, blue and black.

History has shown time and again that the world, seen differently, becomes a different world. *#thedress* takes this full circle: the world we each inhabit becomes the world we see. Like two facing mirrors, we are what we see and we see what we are, an infinite reflection, as far as the eye can see …

POSTSCRIPT:
2021

April 2020, South Downs National Park, UK

I am forty-five minutes into my daily walk in the rolling hills behind my home. My course differs each day, variations on a rural theme of farmland dotted with woods and patches of 'rewilded' scrub. Black-faced lambs hop and scamper back to their mothers with a baa–aa–aa as I pass. A small, silent herd of dairy cows and their calves regard me with doleful eyes as they chew calmly on the fresh, green grass. Skylarks hovering overhead urge me up and on with their relentlessly uplifting trill-trill-trill-eeeee. The occasional raptor circles above, menacing. Very occasionally I meet another walker coming the other way. The usual pleasantries are non-existent; they veer to the far side of the path and look away, avoiding my eye.

The sky is as blue as I've ever seen it: a deep turquoise that seems to crackle with life. The jet trails that usually crisscross the sky from here to Gatwick Airport 20 miles away are absent.

At the top of the hill, I take in the view to the north and see details I've never noticed before. The undulations of distant hills, a steeple here, another there, a silhouetted cluster of high-rise buildings on the horizon, a glint of sunlight off something far away. Everything feels new, different, changed and changing, yet strangely frozen. The chatter of nature does nothing to dispel the sense of deep quiet.

I'm not missed at home, for now. My children are all back from school and university and will be spread through the house, trying to treat their online lessons as 'normal' for the foreseeable future. Our nuclear family, and just our family, is together for every meal. This has been a novelty in recent years; now it's all too routine.

My senses feel heightened, and yet strangely dulled. The air is full of the sounds of nature but feels silent; the atmosphere is clear and crisp yet feels heavy. My surroundings are eminently peaceful, but I feel deeply unsettled: exhilarated by nature's spring commitment to the season ahead but pensive about a future for myself and my family that is entirely unpredictable.

November 2020

Full circle. Trees are leafless again, and again our lives have turned upside down, or more accurately outside in, as we endure a second period of staying home to save lives. A million people around the world have died, and many more, tragically, will follow. There is hope on the horizon with positive outcomes in several vaccine trials, but it will be many months before these will be universally available, allowing life – one hopes – to get back to 'normal' …

The virus known as Severe Acute Respiratory Syndrome Coronavirus 2, SARS-CoV2 (coronavirus) and the disease it causes, COVID-19 (Covid), have devastated millions of families, national health care systems and global economies. The Covid pandemic has also wreaked havoc on the human senses, for the millions of people who have been infected with the virus and for billions more affected by the measures taken to combat its spread.

In early February 2020, two months after Covid was discovered in China, reports emerged that people infected with the coronavirus often lost their ability to smell and taste. This was added to the list of official symptoms by the World Health Organisation and US agencies in April, and by UK agencies a month later. Estimates range up to 88 per cent of cases reporting these symptoms, and while most patients report smell and taste returning to normal within a couple of weeks, around 10 per cent of people fail to regain their senses after a month, and a small number report continuing to experience changes months later. Some experience a continued absence, while for others it's changes to smell or taste functions, be it heightened sensitivity, ordinarily pleasant tastes and odours turning nasty, or perceptions of horrible phantom smells when none are present.

Respiratory viruses like the common cold can cause a temporary loss of smell due to blocked or inflamed nasal passages, and this can also cause a loss of taste. By contrast, coronavirus seems to act directly on the sensory systems that form taste and smell. These are the olfactory (smelling) system, the gustatory (tasting) system and the somatosensory system, an aspect of touch that responds to direct contact with certain chemicals in a process called chemesthesis. Chemesthesis perceives sensations like the heat from chillies (caused by the chemical capsaicin) and the cold of toothpaste (menthol). It occurs mostly in the mouth, but can also take place in other areas[1] and explains the burn of rubbing your eye after cutting up a chilli and the effect of muscle-pain-relieving heat or freezing gels.

Covid disrupts either the neurons in the nose, mouth and elsewhere themselves, or the mechanisms that support them, by blocking some or all of the signals they send to the brain. Like visual neurons, taste and smell neurons build perception by putting together individual neural reactions to specific chemical cues – remember the poor cat we met in Chapter 1 whose brain cell reacted to a diagonal line? A perceived smell or taste is like the music of an orchestra with dozens of different chemical instruments playing designated parts to create a harmonious whole. When Covid disrupts some of the 'instruments', the resulting smell or taste can be like the sound of an orchestra tuning up before a performance.

A particularly strange and disturbing aspect of this sensory disruption is that the experience of disrupted smell and taste often coincides with feelings of anxiety and depressed mood. Strangely, this happens more frequently than with much more dangerous Covid symptoms like high fever and shortness of breath.[2] The researchers who discovered this connection link it to reported neurological Covid symptoms like anxiety, agitation, confusion and seizures. They hypothesise that it might be evidence of coronavirus entering the central nervous system directly via the sensory neurons in the nose and mouth.

In rarer instances, Covid can affect the other senses. A handful of cases of sudden hearing loss after Covid have been reported,[3] and Covid can cause viral conjunctivitis, an infection of the eye.[4] There have also been reported cases of blurred vision while suffering from the disease – including one notorious instance in which the UK prime minister's most senior advisor broke lockdown rules to drive 60 miles with his family in the car in order to 'test his eyesight'. While blurred vision may be an occasional symptom of Covid, this particular account has been widely dismissed.

Beyond those who have been infected with the coronavirus* are the hundreds of millions whose daily sensory experiences have been transformed by the pandemic. Social distancing, personal protective equipment, self-isolation, lockdowns, staycations, support bubbles, furloughs, working from home, quarantines and increased hand hygiene are some of the actions people have been asked to take to suppress the transmission of coronavirus. In addition to introducing our ears to a new vocabulary, these measures have radically altered everyday practices and our relationships with each other, the environment around us, and ourselves. In doing so they have deprived us of sensory needs so basic they were barely acknowledged until they were absent. Seeing other people. Reading the expression on someone's face. Touching the hand or cheek of a loved one. Hugging a friend or colleague. Shaking hands in greeting or agreement. Sharing a space with well-meaning strangers. Singing and dancing in public and private. Gathering in numbers to celebrate or to mourn. Covid measures have made the natural feel alien and frightening, and have forced us to change the habits of a lifetime and try to adopt new ones.

People's day-to-day experience of the Covid pandemic has varied enormously. Thousands are bereaved or continue to suffer symptoms of 'long Covid', while others have had no direct experience of the virus at all. The people we now call 'key workers' have been through an incredibly demanding period, while others have been in a state of suspended animation, furloughed from regular jobs. Some industries have boomed while others have faced enormous hardship, just as some

---

* Approximately 55 million confirmed cases worldwide at the time of writing, but there are probably many times that number of cases that were not formally diagnosed.

individuals have fallen between the cracks of different forms of government support.

In contrast with these stark differences in experience, Covid has had a universal impact on our senses, especially touch. As our most ubiquitous sense – we 'feel' through almost every part of our bodies, both inside and out – touch is also taken most for granted. Pre-Covid would you consciously pause before opening a door, picking up an object in a shop, or sitting next to someone on a train? Covid censures touch, bringing a consciousness to our myriad tactile experiences that is disconcerting and dislocating. Heightening our awareness of own bodies within our immediate environment separates us from that environment, contributing further to feelings of dislocation and isolation.[5]

Some people have found the restrictions of social distancing and PPE and the consequent elimination of skin-to-skin touch a relief. Former US President Trump reportedly remarked that 'maybe this Covid is a good thing … if it means I don't have to shake hands with these disgusting people'.[6] But for hundreds of thousands of others who have not been able to hold the hand of a dying loved one, or who haven't hugged vulnerable parents or spouses or friends in months, or who have had to isolate themselves at home beyond the reach of direct human contact, it has been agony. We heard about 'skin hunger' in Chapter 15, and this sad condition is a widely recognised secondary outcome of Covid and its restrictions, with dire emotional, and mental and physical health consequences including, ironically, depressed immunity to viruses.[7] Sales of furry pets have skyrocketed, and popular magazines suggest techniques for 'solo intimacy' to combat the condition, but as long as restrictions continue, there is little relief to be had.

For all their downsides, it is unarguable that screens have been a godsend during the pandemic for those lucky enough to have access to them, and we have certainly embraced their offerings. Global internet traffic surged by 40 per cent between February and mid April 2020 according to the International Energy Agency.[8] The number of users on Zoom and Houseparty went from hundreds of thousands in early 2020 to millions just a few months later (and ushered in the awkward sensory phenomenon of seeing one's own image on screen while in a work meeting or socialising). Weekly video calls doubled overall and tripled among the over 65s.[9]

Our screens have kept the world turning by allowing us to stay in touch, to continue to work and learn, to socialise, shop and carry on most of the other aspects of our lives that were already gradually moving online. The trends discussed in Chapter 14 have all been accelerated by Covid, and the pandemic may deliver the ultimate triumph of the screen. This inevitably has its own consequences: both the benefits and risks of the dominance of screens in our lives that were becoming apparent before the Covid pandemic have been accelerated and exaggerated since, supplemented with fears of a new myopia epidemic caused by excessive screen use.[10]

It's impossible to know when the Covid pandemic will end. The world is focussing on the roll out of vaccines in 2021; there is no Plan B. Assuming we can eradicate or permanently contain the disease, will we embrace a 'new normal', or fall gratefully back into the lives we stepped out of early in 2020? Will our relationships, our emotional health and our economies recover from the ferocious blow Covid has dealt? Will we reconnect with our senses as they stood in early 2020 or be forever changed? Will future histories identify the extraordinary year of 2020 as a turning point in the human story, or just one of the many bumps along the road?

To all of this, the only answer can be: we'll see.

NOTES

Foreword

1 Jones, C., *Paris: Biography of a City* (London: Penguin Books, 2006).

Prologue

1 The original Tumblr post is no longer available but the image known as #thedress can be seen at www.en.wikipedia.org/wiki/The_dress.

Chapter 1

1 Livingstone, M., *What Art Can Tell Us About the Brain*, lecture recorded at University of Michigan, 2015, accessed at www.youtube.com/watch?v=338GgSbZUYU, 13 August 2016.
2 Livingstone, 2015, p.218.
3 Chau, H.F., Boland, J.E., Nisbett, R.E., Cultural variation in eye movements during perception, Proceedings of the National Academy of Sciences, USA, 2005, Vol.102, No.35, pp.12, 629–33.
4 Livingstone, M., *Vision and Art: The Biology of Seeing* (New York, NY: Abrams, 2014, Second Edition).

5　See the video of this and similar experiments at
　　www.theinvisiblegorilla.com.
6　Mongillo, P. Bono, G., Regolin, L., Marinelli, L., Selective attention
　　to humans in companion dogs, *Canis familiaris, Animal Behaviour*,
　　2010, Vol.80, No.6, pp.1057–63.
7　Livingstone, 2015, ibid.
8　Sugovic, M., Turk, P., Witt, J.K., 'Perceived distance and obesity: it's
　　what you weigh, not what you think', *Acta Psychol* (Amst). March
　　2016; Vol.165: pp.1–8, accessed 6 February 2016.
9　Roberson, D., Davidoff, J., Davies, I.R.L., Shapiro, R.L., 'Colour
　　categories and colour acquisition in Himba and English', in Pitcham,
　　N., Biggam, C.P., *Progress in Colour Studies: Vol. II. Psychological Aspects*
　　(Amsterdam: John Benjamins Publishing, 2006).

Chapter 2

1　Darwin, C., *On the Origin of Species* (First Edition) (London: Murray,
　　1859).
2　Fossil gallery, Burgess Shale website, Royal Ontario Museum,
　　burgess-shale.rom.on.ca/en/fossil-gallery/index.php.
3　www.newscientist.com/article/dn19916-oxygen-crash-led-to-
　　cambrian-mass-extinction.
4　Walcott, C.D., Field Diary Notes, 1909, accessed via Royal
　　Ontario Museum website, burgess-shale.rom.on.ca/en/history/
　　discoveries/02-walcott.php.
5　Zhao, F., Bottler, D.J., Hu, S.Yin, Z., Zhu, M., 'Complexity and
　　diversity of eyes in Early Cambrian ecosystems', *Scientific Reports*,
　　Vol.3, 2013, No.2751.
6　Nilsson, D.-E., 'The functional basis of eye evolution', *Visual
　　Neuroscience*, Vol.30, 2013, accessed at journals.cambridge.org/action/
　　displayFulltext?type=1&fid=8889789&jid=VNS&volumeId=30&
　　issueId=1-2&aid=8889787&bodyId=&membershipNumber=&socie
　　tyETOCSession=, 15 September 2015.
7　Nilsson, D-E., Pegler, S., 'A pessimistic estimate of the time required
　　for an eye to evolve', *Biological Sciences*, 2004, Vol.256, pp.53–58.
8　Halder, G., Callaerts, P., Gehring, W.J., 'Induction of ectopic eyes by
　　targeted expression of the eyeless gene in Drosophila', *Science*, New
　　Series, 1995, Vol.267, No.5205, pp.1788–92.
9　Barinaga, M., 'Focusing on the eyeless gene', *Science*, 1995, Vol.267,
　　pp.1766–67.

10 Gehring, W., Ikeo, K., 'Pax6: mastering eye morphogenesis and eye evolution', *Trends in Genetics*, 1999, Vol.15, No.9.

11 Parker, A., *In the Blink of an Eye: How Vision Sparked the Big Bang of Evolution* (New York, NY: Basic Books, 2003).

12 Lamb, T.D., Collin, S.P., Pugh, E.N., 'Evolution of the vertebrate eye: opsins, photoreceptors, retina and eye cup', *Nature Reviews Neuroscience*, 2007, Vol.8, pp.960–76; Lamb, T.D., 'Evolution of phototransduction, vertebrate photoreceptors and retina', *Progress in Retinal and Eye Research*, 2013, Elsevier, Vol.36, pp.52–119.

13 Xu,. Y., Zhu, S-W., Li, Q-W., 'Lamprey: a model for vertebrate evolutionary research', *Zoological Research*, 2016, Vol.37, No.5, pp.263–69.

14 Nikitina N., Marianne Bronner-Fraser M., Sauka-Spengler T., 'Sea Lamprey Petromyzon marinus: a model for evolutionary and developmental biology', *Cold Spring Harbor Protocols*, 2009.

15 *Blood Lake: Attack of the Killer Lampreys*, www.imdb.com/title/tt3723790.

16 Lamb, 2007, ibid.

17 Banks, M.S., Sprague, E.W., Schmoll, J., Parnell, J.A.Q. Love, G.D., 'Why do animal eyes have pupils of different shapes?', *Science Advances*, 2015, Vol.1, No. 7.

18 Bowmaker, J.K., 'Evolution of colour vision in vertebrates'. *Eye*, 1998, Vol.12 (3b): pp.541–47.

19 Ross, C.F., Kirk, E.C., 'Evolution of eye size and shape in primates', *Journal of Human Evolution*, 2007, Vol.52.

20 Tomasello, M., Hare, B. Lehmann, H., Call, J., 'Reliance on head versus eyes in the gaze following of great apes and human infants: the cooperative eye hypothesis', *Journal of Human Evolution*, 2007, Vol.52, pp.314–20.

21 Tomasello et al., ibid.

Chapter 3

1 Frazer, J.G., *Myths of the Origin of Fire: An Essay* (London: Macmillan and Co. Ltd, 1930).

2 OED, accessed 13 September 2017 at www.oed.com/view/Entry/47 317?redirectedFrom=darkness&

3 Ekirch, A.R., *At Day's Close: A History of Nighttime* (London: Phoenix, 2005).

4 Frazer, J.G., 1930.

5 Darwin, C., 1871, *Descent of Man*, p.45.
6 Pausas, J.G., Keeley, J.E., 'A burning story: the role of fire in the history of life', *BioScience*, 2009, Vol.59, No.7 pp.593–601, accessed at www.biosciencemag.org.
7 Wrangham, R., *Catching Fire: How Cooking Made us Human* (New York, NY: Basic Books, 2009).
8 Gani, M.R. and Gani N.D.S., *Geotimes*, 2008, Vol.1, accessed at www.geotimes.org/jan08/article.html?id=feature_evolution.html, May 2016.
9 Gani and Gani, 2008.
10 NASA map of cumulative lightning strikes, 1995–2013 eoimages. gsfc.nasa.gov/images/imagerecords/85000/85600/lightning_lis_otd_1995-2013_lrg.jpg.
11 Wrangham, 2009, p.3.
12 Goodman, K., McCravy, K.W., 'Pyrophilious insects'; entry in *Encyclopaedia of Entomology*, 2008, pp.3090–93. Capinera J.L. (ed.) (The Netherlands: Springer).
13 Archaeologists have found evidence of butchering forelimbs – a superior meat source normally consumed early by primary hunters – among fossilised remains from 2.6 mya and possibly as long as 3.4 mya. Thompson, J.C. et.al. 'Taphony of fossils from the hominin-bearing deposits at Dikika, Ethiopia', *Journal of Human Evolution*, 2015, Vol.86, pp.112–35; Dominguez-Rodrigo, M. et. Al. 'Cut marked bones from Pliocene archaeological sites at Gona, Afar, Ethiopia, 2005, *Journal of Human Evolution*, Vol.48, pp.109–21.
14 Wrangham, 2009.
15 Wrangham, 2009.
16 Aiello, L., Wheeler, P., 'The expensive-tissue hypothesis: the brain and the digestive system in human and primate evolution', *Current Anthropology*, 1995, Vol.36, pp.199–221.
17 Wrangham, 2009.
18 Burton. F.D., *Fire: The Spark that Ignited Human Evolution* (Albuquerque, NM: UNMP, 2009), p.10.
19 Dunbar, R.I.M, Gowlett, J.A.J. (eds), 'Fireside chat: the impact of fire on hominin socioecology', *From Lucy to Language: The Benchmark Papers* (Oxford: Oxford University Press, 2014).
20 Dunbar, R.I.M., 'The social brain hypothesis and its implications for social evolution', *Annals of Human Biology*, 2009, Vol.36, No.5, pp.562–72.
21 Dunbar, Gowlett, 2014.
22 OED online accessed at www.oed.com/view/Entry/85090?rskey=I FXdH5&result=1&isAdvanced=false#eid.

23 Weissner, P.W., 'Embers of society: firelight talk among the Ju/ Ohansi Bushmen', *Proceedings of the National Academy of Sciences of the United States of America*, 2015, Vol.111 No.39, pp.14027–35.

24 Dunbar, Gowlett, 2014.

25 Sandgathe, D.M., Dibble, H.L., Goldberg, P., McPherron, S.P., Turq, A., Niven, L., & Hodgkins, J., 'Timing of the appearance of habitual fire use', *Proceedings of the National Academy of Sciences of the United States of America*, 2011, Vol.108, No.29.

26 Burton (2009), Wrangham (2009), Dunbar (2009), Dunbar and Gowlett (2014), Aiello and Wheeler (1995), Weissner (2015), Clark, J.D., Harris, J.W.K. (1985) 'Fire and its roles in early hominid lifeways', *African Archaeological Review*, Vol.3, No.1, pp.3–27, Springer.

27 Roebroeks, W., & Villa, P., 'On the earliest evidence for habitual use of fire in Europe', *Proceedings of the National Academy of Sciences of the United States of America*, 2011, Vol.108, No.13, pp.5209–14, doi. org/10.1073/pnas.1018116108; Shimelmitz, R., Kuhn, S.L., Jelinek, A.J., Ronen, A., Clark, A.E., Weinstein-Evron, M., 'Fire at will: the emergence of habitual fire use 350,000 years ago', *Journal of Human Evolution*, 2014, Vol.77, pp.196–203.

28 Gowlett, J.A.J., Wrangham, R.W., 'Earliest fire in Africa: towards the convergence of archaeological evidence and the cooking hypothesis', *Azania: Archaeological Research in Africa*, 2013, Vol.48, No.1, pp.5–30.

29 Burton (2009), Weissner, P.W. (2015), 'Embers of society: firelight talk among the Ju/Ohansi Bushmen', *Proceedings of the National Academy of Sciences of the United States of America*, Vol.111 No.39, pp.14027–35.

30 Dunbar, Gowlett, 2014.

Chapter 4

1 This account is based on the book written by the cave's discoverers, Brunel, Elliette, Chauvet, Jean-Marie, Hillaire, Christian, *The Discovery of the Chauvet-Pont d'Arc Cave* (Saint-Remy-de-Provence: Editions Equinoxe, 2014), as well as observations taken from the documentary *Cave of Forgotten Dreams*, Werner Herzog (Dir.), 2017, viewed at www.youtu.be/dlIEfNbcz7g.

2 Chauvet, J.-M., Brunel Deschamps, E, Hillaire, C. *Chauvet Cave: The Discovery of the World's Oldest Paintings* (London: Thames and Hudson, 1996).

3 Dunbar, R.I.M., 'The social brain: mind, language, and society in evolutionary perspective', *Annual Review of Anthropology*, 2003, Vol.32. pp.163–81.

4 Bruner, E., Lozano, M., 'Extended mind and visua-spatial integration: three hands for the Neanderthal lineage', *Journal of Anthropological Science*, 2014, Vol. 92, pp.273–80.

5 Morriss-Kay, 2010.

6 Guthrie, R.D., *The Nature of Paleolithic Art* (Chicago, IL: Chicago University Press, 2006).

7 Clottes, J. *Cave Art* (London: Phaidon, 2008).

8 Hodgson, D., Watson, B., 'The Visual Brain and the early depictions of animals in Europe and South-East Asia', *World Archaeology*, 2015, Vol.47, No. 5, pp.776–91.

9 Hodgson, D., Watson, B., 'The visual brain and the early depictions of animals in Europe and South-East Asia', *World Archaeology*, 2015, Vol.47, No.5, pp.776–91.

10 Shipman, P., *The Invaders: How Humans and Their Dogs Drove Neanderthals to Extinction* (Boston, MA: Harvard University Press, 2005).

Chapter 5

1 Carpenter, E.S., 'The tribal terror of self-awareness, in Hockings', P. (ed.), *Principles of Visual Anthropology* (The Hague: Mouton, 1975), p.455.

2 Ovid (A.S. Kline's version), *Metamorphoses Book III*, accessed at ovid. lib.virginia.edu/trans/Metamorph3.htm, 15 January 2016.

3 Mellaart, J. (19), 'Catal Hoyuk: a Neolithic town in Anatolia', p.27.

4 Cashdan, E.A., 'Egalitarianism among hunters and gatherers', *American Anthropologist*, 1980, 82 1, pp.116–20, Wiley, London. accessed at onlinelibrary.wiley.com/doi/10.1525/ aa.1980.82.1.02a00100/full, 1 February 2016.

5 Anadolu Agency, 'Çatalhöyük excavations reveal gender equality in ancient settled life', *hurriyetdailynews.com*, citing interview with Ian Hodder, 2014.

6 Hodder, I., 'Catalhoyuk: the leopard changes its spots. A summary of recent work'. *Anatolian Studies*, 2014, Vol.64, pp.1–22.

7 Carpenter, E.S., 'The tribal terror of self-awareness, in Hockings', P. (ed.), *Principles of Visual Anthropology* (The Hague: Mouton, 1975).

8 Prins, H.E.L., Bishop, J., 'Edmund Carpenter: explorations in media and anthropology', *Visual Anthropology Review*, 2002, Vol.17, No.2, pp.110–31.

9 Carpenter, E.S., 'The tribal terror of self-awareness, in Hockings', P. (ed.), *Principles of Visual Anthropology* (The Hague: Mouton, 1975).

10 www.freud.org.uk/2018/07/23/self-reflection-mirrors-in-sigmund-freuds-collection.

11 Chandler, J., 'Little common ground as land grab splits people', *Sydney Morning Herald*, Australia, 15 October 2011, accessed at www.smh.com.au/world/little-common-ground-as-land-grab-splits-a-people-20111014-1lp09.html, 1 February 2016.

12 Carpenter, E.S., *Oh, What a Blow That Phantom Gave Me!* (New York, NY: Holt, Rinehart and Winston, 1972), p.129.

13 Bianchi, R.S., 'Reflections in the sky's eyes', *Notes on the History of Science, 2005*, Vol.4, pp.10–8, Figs. 1–3.

14 Laërtius, Diogenes, *Lives of Philosophers*, II, 33, quoted in Sinisgalli, Rocco, *Perspective in the Visual Culture of Classical Antiquity* (Cambridge: Cambridge University Press, 2012).

15 Seneca, *Naturales Quaestiones*, I, 17, 4, quoted in Sinisgalli, ibid.

16 Toohey, P., *Melancholy, Love and Time*, Chapter 8 (Michigan, MI: University of Michigan Press, 2004).

17 Diener, E. and Wallbom, M., 'Effects of self awareness on antinormative behaviour', *Journal of Research in Personality*, 1976, Vol.10, pp.107–11.

18 Beaman, A.L., Klentz, B., Diener, E., Svanum, S., 'Self-awareness and transgression in children', *Journal of Personality and Social Psychology*, 1979, Vol.37, No.10, pp.1835–46.

19 Cooley, C., *Human Nature and the Social Order*, 1902.

20 Gallup, G.G. Jr, 'Self awareness and emergence of mind in primates', *American Journal of Primatology*, 1982, Vol.2, pp.237–48.

21 Taylor Parker, S., Mitchell, R.W., Boccia M.L., 'Self awareness in animals and humans', *Developmental Perspectives* (Cambridge: Cambridge University Press, 1994).

22 Mortimer, Ian, 'The mirror effect: how the rise of mirrors in the fifteenth century shaped our idea of the individual', *Lapham's Quarterly*, 2016, accessed at www.laphamsquarterly.org/roundtable/mirror-effect.

23 Hockey, D., *Secret Knowledge: Rediscovering the Lost Techniques of the Old Masters* (New York, NY: Viking Studio, 2006).

24 fandomania.com/tv-review-mythbusters-8-27-presidents-challenge

25 Pohl, A., McGuire, J., Toobie, A., 'P6 9 laser diode another day', *Journal of Physics Special Topics*, 2013, accessed at journals.le.ac.uk/ojs1/index.php/pst/article/download/2153/2057.

26 *Hist. Cienc. Saude-Manguinhos*, 2006, Vol.3 suppl. 0, Rio de Janeiro.

27 Alexandra A., 'Reflections from the tomb: mirrors as grave goods in late classical and Hellenistic Tarquinia', *Etruscan Studies*, 2008, Vol.11, No.1, accessed at scholarworks.umass.edu/etruscan_studies/vol11/iss1/1.

28 Harkness, D.E., 'Alchemy and eschatology: exploring the connections between John Dee and Sir Isaac Newton', in *Force*, 1999; J.E., Popkin, R.H. eds 'Newton and religion, Context nature and influence', *Springer-Science and Business Media*, 1999.

29 www.latin-dictionary.net/search/latin/mirare, accessed 1 February 2016.

Chapter 6

1 Schmandt-Besserat, D., Writing systems. [Online]. In D. Pearsall (ed.). *Encyclopedia of Archaeology*, 2008, Oxford, United Kingdom: Elsevier Science & Technology. Available from: 0-search.credoreference.com.wam.city.ac.uk/content/entry/estarch/writing_systems/0, accessed 15 March 2016.

2 Lapidus, I.M., Cities and Societies: A Comparative Study of the Emergence of Urban Civilisation in Mesopotamia and Greece, *Journal of Urban History*, 1986, Vol.12, No.3, pp.257–92, Sage Publications. Accessed 10 March 2016.

3 Englund, R.K., The Ur III Collection of the CMAA, *Cuneiform Digital Library Journal* 2002:1 cdli.ucla.edu/pubs/cdlj/2002/001.html © Cuneiform Digital Library Initiative.

4 Powell, B. *Writing: Theory and History of the Technology of Civilisation* (Hoboken, NJ: Wiley, 2012).

5 Damerow, P., 'The Origins of Writing as a Problem of Historical Epistemology', *Max Planck Institute for the History of Science*, 2006, Berlin.

6 Leeming, D., 'Flood' entry, *The Oxford Companion to World Mythology* (Oxford: Oxford University Press, 2004), p.138.

7 blog.britishmuseum.org/who-was-ashurbanipal/?_ga=2.217768542.950196694.1540219698-2130113507.1539634091.

8 Oed.com, accessed 21 October 2018.

9   www.independent.co.uk/news/uk/this-britain/the-big-
    question-what-is-the-rosetta-stone-and-should-britain-return-it-
    to-egypt-1836610.html, retrieved 21 October 2018.

10 Robinson, A., *The Last Man Who Knew Everything: Thomas Young*
    (Oxford: Oneworld, 2006).

11 Young, T., An Account of Some Recent Discoveries in
    Hieroglyphical Literature, and Egyptian Antiquities (London: John
    Murray, 1823) pp. xiv–xv.

12 Robinson, A., 'Thomas Young and the Rosetta Stone', *Endeavour*,
    2007, Vol.31 No.2, Elsevier. Accessed at via sciencedirect.com,
    12 March 2016.

13 Adkins, L.A., *Empires of the Plain*, 2004, citing Borger, R. (1975–78)
    from the RAS archive.

14 Quinn, J., *In Search of the Phoenicians* (Oxford: Oxford University
    Press, 2017).

15 'Phoenician' entry, Oxford English Dictionary, accessed at www.oed.
    com, 21 October 2018.

16 'adjab' entry, Oxford English Dictionary, accessed at www.oed.com,
    21 October 2018.

17 Powell, B., *Writing: Theory and History of the Technology of Civilisation*
    (Hoboken, NJ: Wiley, 2012).

18 Powell, B., 'Why was the Greek alphabet invented? The epigraphical
    evidence', *Classical Antiquity*, 1989, Vol.8, No.2, University of
    California Press, p.346, www.jstor/stable/25010912.

19 RGS Archives, Rawlinson, *Personal Adventures*, cited in Adkinds
    (2004).

20 Powell, 1989, p.2.

21 Plato, Phaedrus (trans. B. Jowett), Project Gutenberg eBook #1636,
    accessed at www.gutenberg.org/files/1636/1636-h/1636-h.htm,
    29 October 2018.

22 *The Athenaeum*, No.3515, 9 March 1895, p.314.

Chapter 7

1   Digitised copy available at www.bl.uk/manuscripts/Viewer.
    aspx?ref=harley_ms_585_f130r, accessed 1 November 2018.

2   Herbert, K., *Looking for the Lost Gods of England* (Cambridgeshire:
    Anglo Saxon Books, 1994), pp.36–37, and translation of the Nine
    Herbs Charm accessed at www.heorot.dk/woden-notes.html#en52,
    1 November 2018.

3 Gordon, T. (trans. from the original), 'Tacitus on Germany', 1910, via Project Gutenberg accessed 5 November 2017 at www.gutenberg. org/files/2995/2995.txt.

4 www.oed.com/view/Entry/168900?rskey=ToobKm&result=2#eid.

5 Tacitus, ibid.

6 Letter from Gregory 1 to Abbott Mellitus, Epistola 76, PL77 1215–1216, accessed at sourcebooks.fordham.edu/source/ greg1-mellitus.txt, 17 October 2017.

7 Taylor, J.E., *Christians and the Holy Places* (Oxford: Oxford University Press, 1993), cited at www.independent.co.uk/arts-entertainment/ history-hiding-pagan-places-david-keys-reports-on-research-which-casts-doubt-on-the-authenticity-of-1468786.html.

8 Bonser, W., 'The cult of relics in the Middle Ages', 1962, *Folkore*, Vol.73 No.4, pp.234–56.

9 Mommsen, T.E., 'Petrarch's conception of the "Dark Ages"', *Speculum, 1942*, Vol.17, No.2 (April 1942), pp.226–42, accessed at www.jstor.org/stable/2856364 1 November 2018.

10 Chaochao G. et al., 'Reconciling multiple ice-core volcanic histories: The potential of tree-ring and documentary evidence, 670–730 CE', *Quaternary International*, 2016, Vol.394, pp.180–93.

11 Lester, L.K. (ed.), *Plague and the End of Antiquity* (Cambridge: Cambridge University Press, 2007).

12 Whipps, H., 'How smallpox changed the world', *Live Science*, 2008, accessed at www.livescience.com/7509-smallpox-changed-world. html, 15 October 2017.

13 Eco, U., *Art and Beauty in the Middle Ages* (New Haven, CT: Yale University Press, 1986), p.16.

Chapter 8

1 Baragli, S., *Art of the Fourteenth Century* (Los Angeles, CA: Getty Publications, 2007).

2 Ilardi, V., *Renaissance Vision: from Spectacles to Telescopes*, 2007, American Philosophical Society, p.42.

3 Sines, G., Sakellarakis, Y.A., Lenses in Antiquity, *American Journal of Archaeology*, 1987, Vol.91, No.2, pp.191–96 accessed at www.jstor. org/stable/505216, 7 November 2018.

4 Temple, R., *Crystal Sun: Rediscovering a Lost Technology of the Ancient World* (London: Century, 2000).

5 Greenblatt, S., *The Swerve: How the World became Modern* (New York, NY: Norton Books, 2001).

6   Einhard, *The Life of Charlemagne*, trans., 1880, Turner (New York, NY: S.E. Harper & Brothers, 1880), accessed at Fordham University Medieval Sourcebook, sourcebooks.fordham.edu/basis/einhard. asp#Charlemagne%20Crowned%20Emperor.

7   Saengar, P., *Space Between Words: The Origin of Silent Reading* (Stanford, CA: Stanford University Press, 1997).

8   BBC Radio, 'The Carolingian Renaissance', *In Our Time*, 2006, accessed at www.bbc.co.uk/programmes/p003hydz.

9   Although this is disputed by Bloom, J., *Paper Before Print: The History and Impact of Paper in the Islamic World* (New Haven, CT: Yale University Press, 2001).

10 Wilkinson, E., *Chinese History: A New Manual* (Cambridge, MA: Harvard University Press, 2002).

11 Bloom, ibid.

12 Lindberg, D., *Theories of Vision from Al Kindi to Kepler* (Chicago, IL: University of Chicago Press, 1976), p.60.

13 Smith, A.M., 'Alhacen on refraction: a critical edition, with English translation and commentary, 2010, of Book 7 of Alhacen's "De Aspectibus"', the Medieval Latin version of Ibn al-Haytham's 'Kitāb al-Manāzir'. Vol.2. English Translation, *Transactions of the American Philosophical Society*, Vol.100, No.3, Section 2.

14 Al Kahlili, J., news item, 2009, BBC News, accessed at news.bbc. co.uk/1/hi/sci/tech/7810846.stm.

15 Lefèvre, W., ed., 'Inside the camera obscura: optics and art under the spell of the projected image', 2007, *Max Planck Institute for the History of Science*, Berlin, contribution by Abdelhamid I. Sabra, Alhazen's *Optics in Europe: Some Notes on What It Said and What It Did Not Say*, accessed at www.mpiwg-berlin.mpg.de/preprints/p333.pdf 11 November 2018, paragraph 2.3.

16 Lefèvre, W., ed., 'Inside the camera obscura: optics and art under the spell of the projected image', 2007, Max Planck Institute for the History of Science, Berlin, contribution by Abdelhamid I. Sabra, Alhazen's *Optics in Europe: Some Notes on What It Said and What It Did Not Say*, accessed at www.mpiwg-berlin.mpg.de/preprints/ p333.pdf, 11 November 2018, footnote 3.

17 Lindberg, p.80.

18 Whitehouse, D., *Glass: A Short History*, 2012, British Museum.

19 Suger, 1140, translation of original manuscript accessed at www. learn.columbia.edu/ma/htm/ms/ma_ms_gloss_abbot_sugar.htm.

20 Suger, ibid.

21 www.smithsonianmag.com/smart-news/the-first-nativity-scene-was-created-in-1223-161485505.

22 Schiller, G., *Iconography of Christian Art*, Vol.II, 1972 (English trans. from German), Lund Humphries, London, pp.179–81, figs 622–39.

23 Erickson, C., *The Medieval Vision: Essays in History and Perception* (Oxford: Oxford University Press, 1976), p.36.

24 1257–81, exactly the same period as the works on Optics by Witelo, Pecham and Bacon were published. The three scholars are all believed to have visited the court during that period (Ilardi, p.27).

25 Bacon, R., *Opus Major*, 1267, Introduction and trans. Bridges, J.H. (London: Williams and Norgate, 1900).

26 Frugoni, C., trans. McCuaig, W., *Inventions of the Middle Ages* (London: The Folio Society, 2007).

27 From Friar Bartolomeo da San Cordio, *Ancient Chronicle of the Dominican Monastery of St Catherine in Pisa*, 1313, cited in Frugoni, p.2.

28 From Friar Bartolomeo da San Cordio, *Ancient Chronicle of the Dominican Monastery of St Catherine in Pisa*, 1313, cited in Frugoni, above, p.2.

29 Whitehouse, D., *Glass: A Short History* (Washington, DC: Smithsonian Institutions, 2012), p.45.

30 Ilardi, 2007, p.9.

31 Ilardi, 2007, p.51.

32 Holden, B.A. et al., 'Global prevalence of myopia and high myopia and temporal trends from 2000 through 2050', *Ophthalmology*, 2016, Vol.123, No. 5, pp.1036–42, accessed at www.aaojournal.org/article/S0161-6420%2816%2900025-7/fulltext, 5 October 2018.

33 Gombrich, E.H., *The Story of Art* (London: Phaidon, 1950).

34 Gombrich, p.152.

35 For a detailed analysis of the light techniques of Giotto and other early Renaissance Italian painting, see Hills, P., *The Light of Early Italian Painting* (New Haven, CT, and London: Yale University Press, 1987).

36 Ilardi, p.190.

37 Hills, 1987, p.65, footnote 4.

38 Ilardi, 2007, p.60; also Riva, M.A., Arpa, C., Gioco, M., 'Dante and asthenopia: a modern visual problem described during the Middle Ages', *Eye*, 2014, London, Vol.28, p.498, published online 14 January 2010.

39 Ilardi, p.190.

40 Hockney, David, *Secret Knowledge; Rediscovering the Lost Techniques of the Old Masters* (London: Thames and Hudson, 2006), pp.66–67. See also Charles Falco's website at wp.optics.arizona.edu/falco/art-optics/ for information and links on the thesis.

41 For example, the matter was the subject of an entire issue of the journal *Early Science and Medicine*, 2005, Vol. 10, No. 2, 'Optics, instruments and painting, 1420–1720: reflections on the Hockney–Falco thesis'.

42 Steadman, P., *Vermeer's Camera* (Oxford: Oxford University Press, 2001).

43 Falco, C., 'Optics and renaissance art', in *Optics in Our Time*, 2006.

Chapter 9

1 Hughes, B., 2017, *Istanbul: A Tale of Three Cities* (London: Weidenfeld and Nicholson, 2017).

2 Helmasperger Notarial Instrument, 1455, accessed at www.gutenbergdigital.de/gudi/eframes/index.htm, 6 April 2018.

3 Some sources cite 158 copies, others 180.

4 Defoe, D., *History of the Devil* (Boston, MA: CD Strong, 1726).

5 Magno, A.M., *Bound in Venice. The Serene Republic and the Dawn of the Book* (New York, NY: Europa Editions, 2013) pp.27–28.

6 Helmasperger Notarial Instrument, 1455, accessed at www.gutenbergdigital.de/gudi/eframes/index.htm, 6 April 2018.

7 DeFoe, 1726, ibid.

8 Meltzner, M., *The Printing Press*, accessed via Questia at www.questia.com/read/122746489/the-printing-press.

9 Eisenstein, E.L., *The Printing Revolution in Early Modern Europe* (Cambridge: Cambridge University Press, 1983), p.13.

10 www.history.com/news/ask-history/what-is-the-origin-of-the-handshake.

11 1229 Decree of the Council of Toulouse, Canon 14: We prohibit also that the laity should be permitted to have the books of the Old and the New Testament; unless anyone from the motives of devotion should wish to have the Psalter or the Breviary for divine offices or the hours of the blessed Virgin; but we most strictly forbid their having any translation of these books. Accessed at en.wikipedia.org/wiki/Council_of_Toulouse#cite_note-9.

12 Carothers, J.C., 'Psychiatry and the Written Word', *Psychiatry*, 1959, pp.304–18.

13 McLuhan, M., *The Gutenberg Galaxy: The Making of Typographical Man* (Toronto: University of Toronto Press, 1962).

14 Ong, W., *Orality and Literacy: The Technologizing of the Word* (Abingdon: Routledge, 1982, Thirtieth Anniversary Edition), p.117.

15 Act II, Scene ii, Line 535. 'Follow him friends, we'll hear a play tomorrow'.

16 British Shakespeare scholar Gabriel Egan analysed English literary texts written by authors alive between 1550 and 1650 and found that variations on 'see a play' were used in 92 per cent of cases, while variants of 'hear a play' were used only 8 per cent of the time (gabrielegan.com/publications/Egan2001k.htm, accessed 27 March 2018).

17 Walsh, J.J., *The World's Debt to the Irish* (Tradibooks, 1926, 2010 edition).

18 British Library blog accessed at blogs.bl.uk/digitisedmanu-scripts/2017/02/old-english-spell-books.html.

19 These three subjects formed the Trivium, the first phase of education, which was followed at the next stage by the Quadrivium, which consisted of arithmetic, geometry, music and astronomy.

20 'Pythagorus', *Stanford Encyclopaedia of Philosophy* (2005); 'Socrates', *Internet Encyclopedia of Philosophy*, accessed at www.iep.utm.edu/socrates; 'Jesus Christ', *Ancient History Encyclopedia*, 2013 www.ancient.eu/Jesus_Christ.

21 Ong, W., *Orality and Literacy: The Technologizing of the Word* (Abingdon: Routledge, 1982, Thirtieth Anniversary Edition).

22 Camille, M., *Mirror in Parchment: The Luttrell Psalter and the Making of Medieval England* (Chicago, IL: University of Chicago Press, 1997).

23 Camille, M., *Mirror in Parchment: The Luttrell Psalter and the Making of Medieval England* (Chicago, IL: University of Chicago Press, 1997).

24 Bewernick, H., *The Storyteller's Memory Palace: A Method of Interpretation Based on the Function of Memory Systems in Literature* (Bern: Peter Lang AG, 2010), p.39.

25 Yates, F.A., *The Art of Memory* (London: Routledge, 1999).

26 Yates, 1999, pp.20–22.

27 Baxandall, M., *Painting and Experience in Renaissance Italy* (Oxford: Oxford University Press, 1972), p.47.

28 Foer, Joshua, *Moonwalking with Einstein: The Art and Science of Remembering Everything* (London: Penguin, 2011).

29 *Jeremy Norman's History of Information*, accessed at www.historyofin-formation.com/expanded.php?era=1450.

30 Translation of memo by Martin Bormann accessed at en.wikipedia. org/wiki/Antiqua–Fraktur_dispute.

31 Ivins, W., *Prints and Visual Communication* (Boston, MA: MIT Press, 1953).

32 Alexander [ed.] *The Painted Page: Italian Renaissance Book Illumination 1450–1550* (London: Royal Academy of Arts, 1994).

33 Blunt & Raphael, *The Illustrated Herbal* (London: Frances Lincoln Ltd, 1979), pp.113–14.

34 www.historyofinformation.com.

35 Hirsch, *Printing, Selling and Reading 1450–1550* (Wiesbaden: Otto Harrossowitz, 1967), accessed at www.historyofinformation.com/expanded.php?id=3691.

36 Hirsch, *Printing, Selling and Reading 1450–1550* (Wiesbaden: Otto Harrossowitz, 1967), accessed at www.historyofinformation.com/expanded.php?id=3691.

37 Eisenstein, 1983, p.65.

38 Rouse & Rouse, 'Backgrounds to Print: Aspects of the Manuscript Book in Northern Europe of the Fifteenth Century', *Authentic Witnesses: Approaches to Medieval Texts and Manuscripts*, 1991, pp.465–66.

39 Uhlendorf, A., *The Invention of Printing and its Spread Until 1470* (Michigan, MI: Michigan University Press, 1932).

40 Eisenstein, E., *Divine Art: Infernal Machine* (Philadelphia, PA: University of Pennsylvania Press, 2011).

41 Parkes, M.B., Introduction to Peter Ganz (ed.) *The Role of the Book in Medieval Culture* (Turnhout: Brepols, 1986), pp.15–16.

42 Harford, T., '50 things that made the modern economy', BBC World Service, 2017, accessed at www.bbc.co.uk/news/business-41582244 2 April 2018.

43 Harford, ibid.

44 Eisenstein, E.,1983.

45 Oxford English Dictionary, via eInformation in Library app.

46 Greilsammer, Myriam, 'The midwife, the priest, and the physician: the subjugation of midwives in the low countries at the end of the Middle Ages'. *The Journal of Medieval and Renaissance Studies*, 1991, Vol.21: pp.285–329.

47 Karma Lochie, *Covert Operations: The Medieval Uses of Secrecy* (Philadelphia, PA: University of Pennsylvania Press, 1999), pp.199–201.

48 Eamon, W., *Science and the Secrets of Nature: Books of Secrets in Medieval and Early Modern Culture* (Princeton, NJ: Princeton University Press, 1994).

49 Ibid.

50 Bacon, F., *The Great Instauration* (1620) Lindberg, p.344.

51 Bacon, F., 'New Organon', Lindberg, 1990, p.344.

52 *Eye-Witness Account of Image-Breaking at Antwerp, 21–23 August 1566*, University of Leiden, accessed at dutchrevolt.leiden.edu/english/sources/Pages/15660721.aspx.

53 'The Chronicle of the Grey Friars: Edward VI', in *Chronicle of the Grey Friars of London*, Camden Society Old Series: Vol. 53, ed. J.G. Nichols (London, 1852), pp.53–78. *British History Online*, www.british-history.ac.uk/camden-record-soc/vol53/pp53-78, accessed 5 April 2018.

54 Order by Thomas Crowell as the King's Viceregent in September 1538, accessed at historicaltexts.org.

Chapter 10

1 Accessed online at douglasallchin.net/galileo/library/1616docs.htm

2 Blackwell R.J., *Behind the Scenes at Galileo's Trial* (Notre Dame, IN: University of Notre Dame Press, 2006).

3 de Santillana, G., *The Crime of Galileo*, 1955 (Chicago, IL: University of Chicago Press, pp.312–13).

4 Accessed online at bertie.ccsu.edu/naturesci/cosmology/galileopope.html.

5 For example. Fujino, S. et al., 'Job stress and mental health among permanent night workers', *Journal of Occupational Health*, 2001, Vol.43, pp.301–06.

6 Kronfeld-Schor, N., Dominoni, D., de la Iglesia, H., Levy, O., Herzog, E.D., Dayan, T., Helfrich-Forster, C., Chronobiology by moonlight, 2013, Proc R Soc B.

7 Kronfeld-Schor, 2013.

8 Richardson, R., 'Why do wolves howl?', *slate.com*, 2014, accessed at www.slate.com/blogs/wild_things/2014/04/14/why_do_wolves_howl_wolves_do_not_howl_at_the_moon.html on 14 May 2018 also www.nationalgeographic.org/media/wolves-fact-and-fiction accessed on 14 May 2018.

9 Zimecki, M., 'The lunar cycle on human and animal behaviour and psychology', *Postepy Hig Med Dosw*, 2006, access at www.phmd.pl/api/files/view/1953.pdf.

10 V. Gaffney et al., 'Time and a place: a luni-solar "time-reckoner" from 8th millennium BC Scotland', *Internet Archaeology*, 2013, Vol.34, accessed at intarch.ac.uk/journal/issue34/gaffney_index.html, 14 May 2018.

11 Leverington, D., *Babylon to Voyager and Beyond: A History of Planetary Astronomy* (Cambridge: Cambridge University Press, 2003) p.3.

12 Asimov, I., *Eyes on the Universe: A History of the Telescope* (London: Andre Deutch, 1976).

13 Heath, Sir T. (trans.), *Aristarchus of Samos* (Oxford: Clarendon Press, 1913), p.302. Cited in Gingerich, O., 'Did Copernicus Owe a Debt to Aristarchus?', *Journal for the History of Astronomy*, 1985, Vol. 16, No. 1, pp.37–42, accessed at adsabs.harvard.edu/full/1985JHA … 16 … 37G, 16 May 2018. The quote is from Archimedes (*c.*216 BCE), *The Sand Reckoner*.

14 For example, in Joshua 10: 12–13, Joshua commanded the Sun and the Moon to stand still, and they obeyed his command. Psalms 19: 4–6 refers to the Sun moving through the sky, while Chronicles and Psalms both refer to the world (Earth) as standing firm, never to be moved.

15 Chapman, A., *Stargazers: Copernicus, Galileo, the Telescope and the Church* (Oxford: Lion Hudson, 2014), p.34.

16 Rabin, S., 'Nicolaus Copernicus', *The Stanford Encyclopedia of Philosophy*, 2015, Edward N. Zalta (ed.), plato.standford.edu/archives/ fall 2015/entries/Copernicus, accessed 16 May 2018.

17 Chapman, 2014, p.34.

18 Copernicus, N., *Commentariolus*, 1415, trans. Rosen, copernicus. torun.pl/en/archives/astronomical/1/?view=transkrypcja&lang=en accessed via themcclungs.net/physics/download/H/Astronomy/ Commentariolus%20Text.pdf, 16 May 2018.

19 Ibid.

20 Owen Gingerich, 'Did Copernicus owe a debt to Aristarchus?', *Journal for the History of Astronomy*, Vol.16, No.1, February 1985, pp.37–42 suggests he may have come across Aristarchus's ideas, whereas Dava Sobel argues that these works were not available in Latin at the time in Sobel, D., *A More Perfect Heaven: How Copernicus Revolutionized the Cosmos* (New York, NY: Walker & Company, 2012) accessed at adsabs. harvard.edu/full/1985JHA…16…37G.

21 Swerdlow, N M., 'Copernicus's derivation of the heliocentric theory from Regiomontanus's eccentric models of the second inequality of the superior and inferior planets', *Journal for the History of Astronomy*,

2017, Vol. 48, pp.33–61, accessed at journals.sagepub.com/doi/abs/10.1177/0021828617691203, 17 May 2018.

22 Bible, Book of Joshua 10:12.

23 This phrase is widely listed as a quotable quote but no direct reference could be found.

24 Chapman, 2014, p.49.

25 Chapman, 2014, p.50.

26 Full text of Copernicus, N., *de Revolutionibus*, 1543, Preface by Andreas Osiander, accessed at www.geo.utexas.edu/courses/302d/Fall_2011/Full%20text%20-%20Nicholas%20Copernicus,%20_De%20Revolutionibus%20%28On%20the%20Revolutions%29,_%201.pdf, 30 May 18.

27 Singham, M., 'The Copernican myths', *Physics Today*, 2007, Vol.60, No.12, p.48.

28 Van Helden, A., 'The invention of the telescope', *Transactions of the American Philosophical Society*, 1977, Vol.67, No.4, pp.1–67.

29 Reeves, E., *Galileo's Glassworks: The Telescope and the Mirror* (Cambridge, MA: Harvard University Press, 2008), p.117, accessed via Questia at www.questia.com/read/119165310/galileo-s-glassworks-the-telescope-and-the-mirror, 15 April 2018.

30 Van Helden, A., 'The invention of the telescope', *Transactions of the American Philosophical Society*, 1977, Vol.67, No.4, pp.1–67.

31 Van Helden, 1977.

32 Bacon, R., *Opus Majus*, 1267, cited in Reeves, E., 2008.

33 Van Helden, 1977, p.19, citing the magi Giambatista Della Porta.

34 The Galileo Project>Chronology>Timeline, galileo.rice.edu/chron/galileo.html

35 Galileo, *The Starry Messenger*, 1610.

36 Galileo, 1610.

37 Dunn, R., *The Telescope: A Short History* (London: National Maritime Museum, 2009), p.30.

38 Chapman, A., 2014, p.165.

39 Galileo, *Letter to the Grand Duchess Christina of Tuscany*, 1615, accessed online at sourcebooks.fordham.edu/mod/galileo-tuscany.asp 12 June 2018.

40 Chapman, 2014, p.174.

41 Holy Congregation for the Index, *Decree of the Holy Congregation for the Index against Copernicanism*, 1616, accessed online at inters.org/decree-against-copernicanism-1616, 12 June 2018.

42 Galileo, G., *The Assayer*, 1623, accessed at web.stanford.edu.ujsabol/certainty/readings/Galileo-Assayer.pdf.

43 Descartes, R., *Discourse on Method*, 1637.

44 Andrade, E.N. da C., 'Galileo', *Notes and Records of the Royal Society of London*, 1964, Vol.19, No.2, pp.120–30, p.128, accessed at www.jstor.org/stable/3519848, 20 May 2018.

45 Diary of Samuel Pepys, accessed at www.pepysdiary.com/diary/1667/05/26, 1 June 2018.

46 www.visioneng.com/resources/history-of-the-microscope.

47 See a photographed copy of *Micrographia* online at archive.org/stream/mobot31753000817897#page/210/mode/2up.

48 Gest, H., 'The discovery of microorganisms by Robert Hooke and Antoni van Leeuwenhoek, Fellows of The Royal Society', *Notes Rec. R. Soc.* 2004, London, Vol.58 pp.187–201.

49 For example, the Jesuit astronomer Riccioli wrote a comprehensive thesis in favour of Tycho Brahe's cosmology in 1651.

50 Jonson, B., *News from the New World Discovered in the Moon*, 1620, play accessible at www.maths.dartmouth.edu.

51 Wilkins, J., *The Discovery of the World in the Moon*, 1638, play accessible at www.gutenberg.org/files/19103/19103-h/19103-h.htm.

52 Hooke, R., *Micrographia, or, Some Physiological Descriptions of Minute Bodies made with Magnifying Glasses, with Observations and Inquiries made Thereon* (London: Royal Society, 1665).

53 www.smithsonianmag.com/science-nature/Galileos-Revolutionary-Vision-Helped-Usher-In-Modern-Astronomy-34545274.

54 www.space.com/18704-who-discovered-uranus.html.

55 www.britannica.com/place/Neptune-planet/Neptunes-discovery.

56 theplanets.org/distances-between-planets.

57 Asimov, 1976, p.90.

58 Asimov, 1976, p.97.

59 NASA website, www.nasa.gov/mission_pages/hubble/science/star-v1.html.

60 NASA website, exoplanets.nasa.gov/the-search-for-life/life-signs.

61 www.space.com/40831-future-mars-rovers-search-alien-life.html.

Chapter 11

1 De Beaune, S., Palaeolithic Lamps and Their Specialization: A Hypothesis. *Current Anthropology*, 1987, Vol.28, No.4, pp.569–77. Retrieved from 0-www.jstor.org.wam.city.ac.uk/stable/2743501. De Beaune, S., White, R., 'Ice Age lamps', *Scientific American*, 1993,

March issue, accessed at www.academia.edu/416951/Ice_Age_ Lamps.

2 Elrasheedy, A, & Schindler, D., 'Illuminating the Past: Exploring the Function of Ancient Lamps', *Near Eastern Archaeology*, 2015, Vol.78, No.1, pp.3–42.

3 National Candles Association website, candles.org/history accessed 16 July 2016.

4 Stowe, H.B., *Pogaunc People: Their Lives and Loves* (New York, NY: Fords, Howard and Hulbert, 1878), p.230, quoted in Brox, p.13.

5 Guild of Scholars website, gofs.co.uk/dynamicpage.aspx?id=84

6 Schivelbusch, W., *Disenchanted Night: The Industrialization of Light in the Nineteenth Century* (Berkeley, CA: University of California Press, 1995), p.7.

7 Ekirch, A.R., *At Day's Close: A History of Nighttime* (London: Phoenix, 2005).

8 Ekirch, p.303.

9 Ekirch, p.302.

10 For example, Weissner, P.W., 'Embers of society: firelight talks amongst the Ju/Hoansi Bushmen', *Proceedings of the National Academy of Sciences*, 2014, Vol.111, No.39, pp.14027–35, accessed at www.pnas. org/cgi/doi/10.1073/pnas.1404212111.

11 Ekirch, p.304.

12 Chadwick, Richard, 'Who were the skywatchers of the ancient near east?', *Bulletin of the Canadian Society for Mesopotamian Studies*, 1988, Iss.39, pp.5–14.

13 Brox, Jane, *Brilliant: The Evolution of Artificial Light* (London: Souvenir Press, 2010).

14 George, A.R., 'The gods Išum and Ḫendursanga: night watchmen and street-lighting in Babylonia', *Journal of Near Eastern Studies*, 2015, Vol.74, No.1.

15 Brox, Jane, p.23.

16 Ekirch, p.129.

17 Ekirch, p.132.

18 Shakespeare, *A Midsummer Night's Dream*, 1600, Act III, Scene 2.

19 According to blind theologian John Hull, whose audio diaries are presented in *Notes on Blindness*, feature documentary, 2016.

20 Ekirch, p.73.

21 royalparks.org/history, accessed on 16 July 16.

22 www.historyoflamps.com.

23 Griffiths, J., *The Third Man: The Life and Times of William Murdoch, Inventor of Gaslight* (London: Andre Deutsch, 1992).

24 Matthews, W., *An Historical Sketch of the Origins and progress of Gas Lighting* (London: Simpkin and Marshall, 1832). Cited in Griffiths, p.250.

25 Griffiths, 1992, p.257.

26 Matthews, W., *An Historical Sketch of the Origin, Progress and Present State of Gas-Lighting* (London: Rowland Hunter, 1827). Cited in *The London Magazine*, 1827, Hunt and Clark, London, Vol.9, p.515.

27 Goldfarb, S.J., 'A Regency gas burner', *Technology and Culture*, 1971, Vol.12, No.3, p.476.

28 Clark, G. *A Farewell to Alms: A Brief Economic History of the World* (Princeton, NJ: Princeton University Press, 2008).

29 Gilbert, W. trans Mottelay, 1893, *De Magnete*, ibid, p.liii.

30 Morse, R., *The Collected Writings of Benjamin Franklin and Friends*, 2004, Wright Center for Innovation in Science Teaching, Tufts University, Medford www.tufts/edu/as/wright_center/personal_pages/bob_m/franklin_electricity_screen.pdf.

31 Priestley, J., ibid, pp.180–81.

32 Brox, p.105.

33 *La Lumiere Electrique*, 1885, cited in Shivelbusch, 1995, p.124.

34 Whipple, F.H., *Municipal Lighting, Detroit*, 1888, p.157, cited in Schivelbusch, p.127.

35 Wells, H.G., *When the Sleeper Wakes*, 1910, cited in Shivelbusch.

36 Stevenson, R.L., A plea for Gas Lamps, *Virginibus Puerisque and Other Papers* (New York, NY: Charles Scribner's Sons, 1893), pp.227–78, cited in Brox, p.104.

37 Howell, J.W., Schroeder, H., 1927, pp.30–35.

38 Spear, B., 'Let there be light! Sir Joseph Swan and the incandescent light bulb', *World Patent Information*, 2013, Vol.35, No.1, Elsevier, pp.38–41.

39 McPartland, D.S., *Almost Edison: How William Sawyer and Others Lost the Race to Electrification*, 2006, Doctoral thesis UMI No. 3231946, p.20; accessed via ProQuest on 25 July 2016.

40 Marshall, D., *Recollections of Edison*, 1931, Christopher Howe, Boston, p.60, cited in McPartland, D.S., *Almost Edison: How William Sawyer and Others Lost the Race to Electrification*, 2006, Doctoral Thesis UMI: 3231946 accessed 23 July 2016.

41 McPartland, 2006, p.22.

42 Ibid, p.26.

43 'In the Stores', *New York Herald*, cited in Brox, ibid, p.123.

44 In Britain, for example, see the British Astronomical Association's Campaign for Dark Skies at www.britastro.org/dark-skies and in the USA the International Dark-Sky Association at www.darksky.org.

Chapter 12

1 Boring, E.G., 'Dual role of the Zeitgeist in scientific creativity', *Scientific Monthly*, 1950, 80, pp.101–06.

2 Ford, B.J., 'Scientific Illustration in the Eighteenth Century', Chapter 24 in Porter, R. (ed.), 2003, *The Cambridge History of Science*, Cambridge University Press.

3 Museum of the History of Science, Oxford University, MHS Collection Database Search, accessed at www.mhs.ox.ac.uk/collections/imu-search-page/record-details/?TitInventoryNo=62524&querytype=field&thumbnails=on////////&irn=3704

4 Fiorentini, E., 'Camera obscura vs. camera lucida: distinguishing nineteenth century modes of seeing', 2006, *Max Plank Institute for History of Science*, Vol. 307, p.7, accessed online at www.researchgate.net/publication/41125012_Camera_obscura_vs_camera_lucida_-_distinguishing_early_nineteenth_century_modes_of_seeing, 11 July 2018.

5 Fiorentini, 2006, p.12.

6 Galassi, P., *Before Photography: Painting and the Invention of Photography* (New York, NY: Museum of Modern Art, 1981), p.11.

7 Galassi, etc.

8 Benjamin, W., *The Work of Art in the Age of Mechanical Reproduction* (London: Penguin Great Ideas, 1936).

9 Sperling, J., 'Multiples and reproductions: prints and photographs in nineteenth century England – visual communities, cultures and class', in Kromm J. (ed.) (2010), *A History of Visual Culture*, Berg, Oxford.

10 His biographer, writing in 1902, describes 'a continual recurrence of fits of depression, sometimes lasting for weeks together'. Litchfield, R.B., *Tom Wedgwood: The First Photographer; An account of his life, his discovery and his friendship with Samuel Taylor Coleridge, including the letters of Coleridge to the Wedgwoods and an examination of accounts of alleged earlier photographic discoveries* (London: Duckworth and Co., 1903), p.24, accessed at archive.org/stream/tomwedgwoodfirst00litcrich#page 20 July 2018.

11 Davy, H., 'An account of a method of copying paintings upon glass, and of making profiles, by the agency of light upon nitrate of silver, with observations by Humphrey Davy. Invented by T. Wedgwood, Esq.', *Journals of the Royal Institution*, 1802, Vol. 1, London.

12 Litchfield, 1902, p.198.

13 Litchfield, 1902, notes that Edward Fox-Talbot, eventually a successful pioneer in photography, remarked that a chemist friend of his had been put off by Davy's account, and that he may have been too had he read it before undertaking his own experiments, p.201.

14 Quoted in Buckland, G., *Fox Talbot and the Invention of Photography* (London: Scolar Press, 1980), p.38.

15 Quoted in Buckland, 1980, p.38.

16 Buckland, 1980, p.38.

17 Fox Talbot, H., *The Pencil of Nature*, 1844, Longman, Brown, Green and Longmans, London, accessed via Gutenburg.org at www.gutenberg.org/files/33447/33447-pdf.pdf, 20 July 2018.

18 Fox Talbot, 1844, p.4.

19 Buckland, 1980, p.30, quoting from Talbot's notebook of February 1835.

20 Unknown, 'Fine Arts. The Daguerre Secret', *The Literary Gazette; and Journal of the Belles Lettres, Arts, Sciences, &c,* 1839 (London) No.1179 (Saturday, 24 August 1839), pp.538–39.

21 Wood, D.W., 'The arrival of the Daguerreotype in New York', monograph for The American Photographic Historical Society (New York), 1995, accessed at www.midley.co.uk/daguerreotype/newyork.htm, 25 July 2018.

22 Unknown Editor, 'New discovery', *Blackwood's Edinburgh Magazine*, 1839, Edinburgh and London, 45:281 (March 1839): pp.382–91. Accessed via Gary W. Ewer, ed., *The Daguerreotype: an Archive of Source Texts, Graphics, and Ephemera*, www.daguerreotypearchive.org

23 Rung, A.M., 'Joseph Saxton, inventor and pioneer photographer', *Pennsylvania History Journal*, 1940, Vol.7, No.3 (July, 1940), pp.153–58, Pennsylvania University Press, accessed at www.jstor.org/stable/27766416.

24 Meredith, R., *Mr. Lincoln's Camera Man: Mathew B. Brady* (New York, NY: Dover, 1974), p.68.

25 Douglass, F., *Narrative of the Life of Frederick Douglass* (Boston, MA: Anti Slavery OFfice, 1845, 2013 Amazon edition), p.5.

26 Douglass, 1845, p.34.

27 Ibid, p.89.

28 Stauffer, J., Trodd, Z., Bernier, C-M., *Picturing Frederick Douglass* (New York, NY: W.W. Norton and Company Inc., 2015, Revised Edition), p.xvi.

29 Stauffer et al., 2015, p.xvi.

30 Douglass, F., 'Lecture on Pictures', 1861.

31 Stauffer, J., Trodd, Z., Bernier, C-M., 2015, p.ix.

32 For example, Baudelaire, Salon of 1859, from Mayne, J. (ed.), *Charles Baudelaire: The Mirror of Art*, 1955, Phaidon Press Limited, London, accessed at www.csus.edu/indiv/o/obriene/art109/readings/11%20 baudelaire%20photography.htm.

33 Stauffer, et al., 2015, p.247.

34 See a long discussion of the various arguments by New York Times writer Errol Morris at opinionator.blogs.nytimes.com/2007/09/25/ which-came-first-the-chicken-or-the-egg-part-one.

35 André Rouillé, A, Lemagny, J.-C., *A History of Photography* (Cambridge: Cambridge University Press, 1988).

36 Ings, S., *The Eye: A Natural History*, 2007 (London: Bloomsbury Publishing, 2007).

37 Maas, J., *Victorian Painters* (London: Barrie and Jenkins, 1978).

38 McCouat, P., 'Early influences of photography on art, part 2', *Journal of Art in Society*, 2012, vaccessed at www.artinsociety.com/ pt-2-photography-as-a-working-aid.html.

39 McCouat, P., 2012.

40 Gossman, L., 'The important influence of paintings in early photograph', referencing his book *Thomas Annan of Glasgow: Pioneer of the Documentary Photograph*, 2015, accessed at brewminate.com/ the-important-influence-of-paintings-in-early-photography.

41 Cromby, I., 'The madonna of the future', *Art Journal*, 2004, 43, National Gallery of Victoria, accessed at www.ngv.vic.gov.au/essay/ the-madonna-of-the-future-o-g-rejlander-and-sassoferrato.

42 Baudelaire, Salon of 1859, from Mayne, J. (ed.), *Charles Baudelaire: The Mirror of Art*, 1955, Phaidon Press Limited, London, accessed at www.csus.edu/indiv/o/obriene/art109/readings/11%20 baudelaire%20photography.htm, 28 November 2018.

43 They were 22in by 18in. Carletonwatkins.org.

44 Victoria and Albert Museum, *Sea and Sky: Photographs by Gustave Le Gray 1856–1857*, 2003, exhibition notes, accessed at www. vam.ac.uk/content/articles/s/gustave-le-grey-exhibition, 3 December 2018.

45 Gombrich, E.H., *The Story of Art* (London: Phaidon, 1989, Fifteenth Edition). The original edition included a Muybridge sequence

showing the movement of a galloping horse as an example of how photography showed artists what the eye couldn't discern.

46 www.telegraph.co.uk/news/2017/11/16/leonardo-da-vincis-salvator-mundi-sells-450-million-342-million, accessed 2 December 2018.

47 See for example de Font-Reaulx, D., *Painting and Photography: 1839–1914* (Paris: Flammarion, 2012).

48 Boucicault, D., *The Octoroon*, 1859, accessed at archive.org/stream/octoroonorlifein00bouc/octoroonorlifein00bouc_djvu.txt referenced in www.phrases.org.uk/meanings/camera-cannot-lie.html, accessed 10 December 2018.

49 www.phrases.org.uk/meanings/camera-cannot-lie.html accessed 10 December 2018.

50 en.wikipedia.org/wiki/A_picture_is_worth_a_thousand_words#cite_note-1.

Chapter 13

1 *Birth of a Nation* film poster accessed at commons.wikimedia.org/w/index.php?search=birth+of+a+nation&title=Special:Search&go=Go&searchToken=qrkxlrnmpsmpozhhwximo2bg#/media/File:Birth_of_a_Nation_-_Academy.jpg.

2 Subtitle from *Birth of a Nation* attributed to Woodrow Wilson, selectively quoted from his five-volume series *History of the American People*; film accessed at ia902702.us.archive.org/29/items/dw_griffith_birth_of_a_nation/birth_of_a_nation_512kb.mp4.

3 D'Ooge, C., *Symposium on 'Birth of a Nation'*, 1994, US Library of Congress, accessed at www.loc.gov/loc/lcib/94/9413/nation.html

4 Rothman, J., 'When bigotry paraded through the streets', *The Atlantic*, December 2016, accessed at www.theatlantic.com/politics/archive/2016/12/second-klan/509468, 29 December 2018.

5 Rossell, D., The Magic Lantern, Published in von Dewitz et al. (ed.), *Ich Sehe was, was du nicht siehst! Sehmaschinen und Bilderwelten* (Cologne: Steidl/MuseumLudwig/Agfa Fotohistoram, 2002), accessed at www.academia.edu/345943/The_Magic_Lantern.

6 Williams, A., *Republic of Images* (Cambridge, MA: Harvard University Press, 1992), p.24.

7 Williams, 1992, p.27.

8 Bowser, E., *The Transformation of Cinema* (Berkeley, CA: University of California Press, 1994); Butsch, W. (2000) 'The making of American

audiences', *International Labor and Working-Class History*, No. 59, Workers and Film: As Subject and Audience (Cambridge: Cambridge University Press, 2001), pp.106–20.

9 Quoted in '"Art [and history] by lightning flash": The birth of a nation and black protest', Roy Rosenzweig Center for History and New Media website, chnm.gmu.edu/episodes/the-birth-of-a-nation-and-black-protest, accessed 7 January 2014.

10 Lekich, J., 'Lillian Gish: First Lady of the silent screen', *Globe and Mail* (Vancouver), 24 October 1986.

11 Benbow, M., 'Birth of a quotation: Woodrow Wilson and "like writing history with lightning"', *The Journal of the Gilded Age and Progressive Era*, 2010, Vol.9, No.4, pp.509–33. Published by Society for Historians of the Gilded Age & Progressive Era.

12 Cook, D.A., Sklar, R. History of the Motion Picture, *Encyclopaedia Britannica*, www.britannica.com/art/history-of-the-motion-picture/ The-silent-years-1910-27, 12 January 2019.

13 Levaco, R., 'Censorship, ideology, and style in Soviet cinema', 1984, *Studies in Comparative Communism*, Vol.XVII, Nos.3 and 4, pp.173–83 accessed at JSTOR.

14 Prince, S. Hensley, W., 'The Kuleshov effect: recreating the classic experiment', *Cinema Journal*, Vol.31, No.2 (Winter, 1992), pp.59–75, accessed at JSTOR.

15 Welch, D., *Propaganda and the German Cinema, 1933–1945* (London: IB Tauris, 1983) pp.12–13.

16 Welch, 1983, p.15.

17 Riefenstahl, L., 1935, cited in Sontag, S., *Fascinating Fascism*, 1975, accessed at docs.google.com/viewer?a=v&pid=sites&srcid=ZGVmY XVsdGRvbWFpbnxmYXNjaXNtYW5kbWFzY3VsaW5pdHl8Z3g 6NDAxMjhmNmEzN2M1OTM3, 3 January 2019.

18 Welch, D., 1983.

19 Hagopian, K.J. (undated), Film notes, New York State Writers' Institute, www.albany.edu/writers-inst/webpages4/filmnotes/ fns07n6.html, and Pierpont, C.R. (2015), review of Wieland, K. (2015) 'Dietrich & Riefenstahl: Hollywood, Berlin, and a Century in Two Lives', Liveright accessed at www.newyorker.com/ magazine/2015/10/19/bombshells-a-critic-at-large-pierpont.

20 Woolley, S., promotional material for his film *Their Finest* about the Film Unit in the Second World War, 2017, based on Lissa Evans' novel *Their Finest Hour and a Half* London: Black Swan, 2010) accessed at www.pressreader.com/uk/daily-express/20170413/28173 2679350947.

21 Richards, J., 'Cinemagoing in worktown: regional film audiences in 1930's Britain', *Historical Journal of Film, Radio and Television*, 1994, Vol.14 No.2, p.147.

22 Kuhn, A., *An Everyday Magic: Cinema and Cultural Memory* (London: IB Tauris, 2002) p.3.

23 Capra, F., *The Name above the Title* (New York, NY: Bantam Books, 1971).

24 Gallese, V., Guerra, M., 'Embodying movies: embodied simulation and film studies', *Cinema*, 2012, Vol.3, pp.183–210, accessed at core. ac.uk/download/pdf/96698686.pdf.

25 See, for example, Hickok, J., 'Eight Problems for the Mirror Neuron Theory of Action Understanding in Monkeys and Humans', *J Cogn Neurosci*, 2009, Vol.21, No.7, pp.1229–43, accessed at www. mitpressjournals.org/doi/full/10.1162/jocn.2009.21189?url_ver=Z39.88-2003&rfr_id=ori%3Arid%3Acrossref.org&rfr_dat=cr_ pub%3Dpubmed.

26 Smith, T.J. et al., 'A window on reality: perceiving edited moving images', *Current Directions in Psychological Science*, 2012, Vol.21, No.2, pp.107–13, accessed at journals.sagepub.com/doi/abs/10.1177/0963 721412437407?journalCode=cdpa.

27 Deleuze, G., *Cinéma I: L'image-mouvement* (Paris: Les Éditions de Minuit, 1983). (Trans. 1986, *Cinema 1: The Movement-Image*).

28 Hasson, U. et al., 'Neurocinematics: The neuroscience of film', *Projections*, 2009, Vol.2, No.1, Summer 2008: pp.1–26.

29 Shaviro, S., *The Cinematic Body* (Minneapolis, MN: University of Minnesota Press, 1993).

30 Early Television Museum, website accessed at www.earlytelevision. org/prewar.html.

31 Number of TV Households in America, 1950–1978, accessed at www.tvhistory.tv/facts-stats.htm, 5 January 2019.

32 Moran, J., *Armchair Nation* (London: Profile Books, 2013).

33 In areas where television was available, web.archive.org/ web/20070831035110/ www.freetv.com.au/Content_Common/ pg-History-of-TV.seo.

34 Turner, H.A. Jnr., *Germany from Partition to Reunification* (New Haven, CT: Yale University Press, 1992), p.98; Kuhn, R., *The Media in France* (Abingdon: Routledge, 1995), p.99.

35 Campbell, et al., *Tuning In, Tuning Out Revisited: A Closer Look at the Causal Links Between Television and Social Capital*, 1999, citing studies from the US, UK, Australia and South Africa, accessed at sites.hks. harvard.edu/fs/pnorris/Acrobat/TVAPSA99.PDF, 15 January 2019.

36 Putnam, R. 'Bowling alone: America's declining social capital', *Journal of Democracy*, Vol.6, No.1, January 1995, pp.65–78.
37 Snowdon, C., *Closing Time: Who's Killing the British Pub?*, report for the Institute of Economic Affairs, 2016, accessed at www.iea.org.uk/ sites/default/files/publications/files/Briefing_Closing%20time_web. pdf.
38 Stephens, M., *The Rise of the Image and the Fall of the Word* (Oxford: Oxford University Press, 1998).
39 Campbell et al., 1999.
40 Putnam, 1995.
41 Ibid.
42 See, for example, Xiong, S. et al., 'Time spent in outdoor activities in relation to myopia prevention and control: a meta-analysis and systematic review', *Acta Ophthalmol., 2017*, Vol.95, No.6, pp.551–66, accessed at www.ncbi.nlm.nih.gov/pmc/articles/PMC5599950.
43 Sherwin, J.C. et al., 'The association between time spent outdoors and myopia in children and adolescents: a systematic review and meta-analysis', *Opthalmology*, 2011, Vol.119, No.10, pp.2141–51.
44 See www.huffingtonpost.com/mark-blumenthal/did_nixon_ win_with_radio_liste_b_729967.html for an interesting analysis of the Kennedy-Nixon TV/Radio debate legend. The author cites a 2002 experiment in which subjects were divided into two groups and either watched a video of the debate or listened to the audio alone. The results were similar to those pollsters found in 1960. (Druckman, J.N., 'The power of television images: the first Kennedy–Nixon debate revisited', *The Journal of Politics*, 2003, Vol.65, No.2, pp.559–71.

Chapter 14

1 www.engadget.com/2007/01/09/live-from-macworld-2007-steve-jobs-keynote.
2 www.engadget.com/2007/07/25/apple-sold-270-000-iphones-in-the-first-30-hours.
3 Newzoo research, accessed at venturebeat.com/2018/09/11/ newzoo-smartphone-users-will-top-3-billion-in-2018-hit-3-8-billion-by-2021.
4 www.zenithmedia.com/smartphone-penetration-reach-66-2018.
5 Power, S. et al., 'Sleepless in school? The social dimensions of young people's bedtime rest and routines', 2017, *Journal of Youth Studies*, accessed via Taylor & Francis. 'One in five young people lose sleep

over social media', *ScienceDaily*, accessed via www.sciencedaily.com/releases/2017/01/170116091419.htm, 16 January 2017.

6 Ofcom, *Communications Market Report*, 2018, accessed at www.ofcom.org.uk/__data/assets/pdf_file/0022/117256/CMR-2018-narrative-report.pdf.

7 Ofcom, *Communications Market Report*, 2018, accessed at www.ofcom.org.uk/__data/assets/pdf_file/0022/117256/CMR-2018-narrative-report.pdf.

8 Boyd, D.M., Ellison, N.B., 'Social network sites: definition, history, and scholarship', *Journal of Computer-Mediated Communication*, 2007, Vol.13, No.1, accessed at onlinelibrary.wiley.com/doi/full/10.1111/j.1083-6101.2007.00393.x?scrollTo=references.

9 MySpace was founded in 2003 and within two years it became one of the US's most visited websites. When Rupert Murdoch's NewsCorp paid $580 million for it two years later it was widely considered a bargain. It was eventually sold to Justin Timberlake for $35 million. In 2005 the British social network site Friends Reunited was bought for £120 million and a couple of years later AOL (beaten out by News on the MySpace deal) paid $850 million for Bebo.

10 Facebook investor relations statistics accessed at zephoria.com/top-15-valuable-facebook-statistics, 18 January 2019.

11 More than 90 per cent of Facebook users access the site from mobile devices, according to blog.bufferapp.com/social-media-trends-2018.

12 Monthly active users of the major sites in January 2019 were: YouTube 1.9 billion; What'sApp 1.5 billion; WeChat (Chinese) and Instragram 1 billion; QQ (Chinese) 800 million; QZone and Douyin (Chinese) 500 million; Sina Weibo (Chinese) 450 million; Reddit and Twitter 330 million; Snapchat 300 million). Source: Statista.com.

13 Bansal, A., Garg, C., Pakhare, A., Gupta, S., 'Selfies: A boon or bane?', *J Family Med Prim Care*, 2018, Vol.7, No.4, pp.828–31.

14 Cascone, S., 'For a project called "selfie harm", the photographer Rankin asked teens to photoshop their own portraits. What they did was scary', *ArtNet News*, 2019, accessed at news.artnet.com/art-world/rankin-selfie-harm-1457959.

15 Twenge, J.M., *Generation Me: Why Today's Young Americans Are More Confident, Assertive, Entitled – And More Miserable Than Ever Before* (New York, NY: Atria Books, 2014); Twenge, J.M., *The Narcissism*

Epidemic: Living in the Age of Entitlement (New York, NY: Atria Books, 2010).

16 See a summary of the debate with references detailed at Pierre, J., 'The Narcissism Epidemic and What We Can Do About It', 2016, at www.psychologytoday.com/gb/blog/psych-unseen/201607/the-narcissism-epidemic-and-what-we-can-do-about-it.

17 Across the whole group, accessing social media was the main reason the students used the internet (98 per cent), higher than research (93 per cent), shopping (85 per cent) or accessing news (81 per cent). Reed, P. et al., 'Visual social media use moderates the relationship between initial problematic internet use and later narcissism', *Open Psychology Journal*, 2019, Vol.12, pp.163–70, accessed at www.benthamopen.com/FULLTEXT/TOPSYJ-11-163.

18 www.cosmopolitan.co.uk/beauty-hair/news/a25314/uk-eyebrow-trends-statistics-2014.

19 www.groominglounge.com/blog/marketing-stuff/men-vs-women-who-spends-more-time-grooming.

20 www.businesswire.com/news/home/20180910005394/en/Global-Male-Grooming-Products-Market-2018-2023.

21 International Society of Aesthetic Plastic Surgeons, *Global Statistics 2017*, www.isaps.org/wp-content/uploads/2018/10/2017-Global-Survey-Press-Release-Demand-for-Cosmetic-Surgery-Procedures-Around-The-World-Continues-To-Skyrocket_2_RW.pdf.

22 Under 30s in the US are nearly three times more likely to sport a tattoo than over 45s, and more than twenty times as likely to have a body piercing. Pew Research Centre, *Millennials: A Portrait of Generation Next, Confident, Connected*, 2010, *Open to Change*, www.pewsocialtrends.org/files/2010/10/millennials-confident-connected-open-to-change.pdf.

23 Sanghvi, R., *Wired Magazine*, 2006, www.wired.com/2016/09/everyone-hated-news-feed-then-it-became-facebooks-most-important-product.

24 Haynes, T., *Dopamine, Smartphones and You: A Battle for your Time*, 2018, sitn.hms.harvard.edu/flash/2018/dopamine-smartphones-battle-time.

25 Zuboff, S., *The Age of Surveillance Capitalism* (London: Profile Books, 2019), pp.452–54.

26 Kegan. R., *The Evolving Self* (Cambridge, MA: Harvard University Press, 1992).

27 McAdams, D., 'Life authorship in emerging adulthood', *The Oxford Handbook of Emerging Adulthood* (Oxford: Oxford University Press, 2015).

28 Layard, R., *Happiness* citing Solnick and Hemenway, 1998 (London: Penguin Books, 2005).

29 Hennigan, K. et al., 'Impact of the introduction of television on crime in the United States: empirical findings and theoretical implications', *Journal of Personality and Social Psychology*, 1982, Vol.42, No.3, pp.461–77; Yong, H. et al., 'Exploring the effects of television viewing on perceived life quality', *Mass Communication and Society*, 2010, Vol.13, No.2, pp.118–38, cited in Zuboff, 2019.

30 Zuboff, 2019, pp.462–65.

31 Twenge, J.M., 'iGen: Why today's super-connected kids are growing up less rebellious, more tolerant, less happy – and completely unprepared for adulthood – and what that means for the rest of us', 2017, presaged in a lengthy *Atlantic* article, www.theatlantic.com/magazine/archive/2017/09/has-the-smartphone-destroyed-a-generation/534198.

32 See links to nine rebuttals at www.digitalnutrition.com.au/5-rebuttals-to-that-smartphone-article-in-the-atlantic.

33 81 per cent said social media makes them feel more connected to what's going on in their friends' lives, with about half of them saying it makes them feel 'a lot' more connected. 70 per cent said social media makes them feel more in touch with their friends' feelings, and a similar number said social media lets them show their creative side. 45 per cent complained about all the 'drama' on social media; 40 per cent said they feel pressure to only post content that makes them look good to, or that will get lots of comments or likes. Around 25 per cent said social media has had a negative impact on their self-esteem, though only 4 per cent said it makes them feel 'a lot' worse about their lives, www.pewinternet.org/2018/11/28/teens-and-their-experiences-on-social-media.

34 Reported at www.theguardian.com/society/2018/nov/22/mental-health-disorders-on-rise-among-children-nhs-figures.

35 Office of National Statistics figures accessed at www.ons.gov.uk/peoplepopulationandcommunity/birthsdeathsandmarriages/deaths/bulletins/suicidesintheunitedkingdom/2017registrations#suicide-patterns-by-age.

36 Morgan, C. et al., 'Incidence, clinical management, and mortality risk following self-harm among children and adolescents: cohort study

in primary care', *BMJ*, 2017, Vol. 359, accessed at www.bmj.com/content/359/bmj.j4351.

37 Mercado, M. et al., 'Trends in emergency department visits for nonfatal self-inflicted injuries among youth aged 10 to 24 years in the United States, 2001–2015', JAMA, 2017, Vol.318, No.19, pp.1931–33, accessed at jamanetwork.com/searchresults?q=mercado&allJournals=1&SearchSourceType=1&exPrm_qqq={!payloadDisMaxQParser%20pf=Tags%20qf=Tags^0.0000001%20payloadFields=Tags%20bf=}%22mercado%22&exPrm_hl.q=mercado, Center for Disease Control analysis of Emergency Dept Admissions.

38 Mojtobaj. R. et al., 'National trends in the prevalence and treatment of depression in adolescents and young adults', *Pediatrics*, 2016, Vol.138, No.6, accessed at pediatrics.aappublications.org/content/143/1?current-issue=y, 20 January 2019.

39 Curtin. S.C. et al., 'Increase in suicide in the United States, 1999–2014', *NCHS Data Brief*, 2016, No.241, pp.1–8, accessed at www.ncbi.nlm.nih.gov/pubmed/27111185, 20 January 2019.

40 Zuboff, 2019, pp.452–54.

41 In 2018 Oxford University researchers found that nearly all of the million odd Android Apps they studied (and there are millions more of them) routinely send user information to Android owner Google. More than 40 per cent send data to Facebook. Binns, R. et al., 'Third party tracking in the mobile ecosystem', 2018, WebSci '18: 10th ACM Conference on Web Science, 27–30 May 2018, Amsterdam, Netherlands.

42 A Quora response to the question 'What does Google know about me' returns the anecdote of a 35-year-old man who had been living with a woman with a son from a previous marriage for a year. He had decided he wanted to marry her. Then one day, as he put it, 'all of Google's ads for wedding rings were replaced with ads for Ashley Madison, porn, and dating sites'. He couldn't understand why the change happened but it got him thinking about what might be wrong with his relationship. He concluded that he didn't actually love his girlfriend, and had strong feelings for another woman who had recently come into his life. He broke it off with his girlfriend and took the other woman out the same night. There could not be a more potent example of Zuboff's claim that Big Tech seeks to predict what we do. Accessed at quora.com (direct URL not available).

43 Subramanian, S., 'Inside the Macedonian fake news complex', *Wired Magazine*, 2017, www.wired.com/2017/02/veles-macedonia-fake-news.

44 Zuboff, S., 2019, ibid.

45 www.marketwatch.com/story/want-to-delete-facebook-read-what-happened-to-these-people-first-2018-07-27.

Chapter 15

1 Ofcom, *Digital Day 2016*, stakeholders.ofcom.org.uk/binaries/research/cross-media/2016/Digital_Day_2016_Overview_of_findings_charts.pdf. Includes email and instant messaging but not the huge volume of messaging that goes through social media platforms.

2 For example: www.forbes.com/sites/neilhowe/2015/07/15/why-millennials-are-texting-more-and-talking-less/#3cde0b4e5576.

3 Selzer, L.J., et al., 'Instant messages vs. speech: hormones and why we still need to hear each other', *Evolution of Human Behavior*, 2012, Vol.33, No.1, pp.42–45, accessed at www.ncbi.nlm.nih.gov/pmc/articles/PMC3277914, 21 January 2019.

4 McGann, J. et al., 'Poor human olfaction is a 19th-century myth', *Science, 2017*, Vol.356, No.6338, accessed at science.sciencemag.org/content/356/6338/eaam7263, 21 January 2019.

5 Lewis, J.G., 'Smells ring bells: How smell triggers memories and emotions: Brain anatomy may explain why some smells conjure vivid memories and emotions', 2015, accessed at www.psychologytoday.com/gb/blog/brain-babble/201501/smells-ring-bells-how-smell-triggers-memories-and-emotions.

6 www.independent.co.uk/news/media/advertising/the-smell-of-commerce-how-companies-use-scents-to-sell-their-products-2338142.html.

7 www.addmaster.co.uk/scentmaster/scentmaster-technology.

8 Kohli, P. et al., 'The association between olfaction and depression: A systematic review', *Chem Senses*, 2016, Vol.41, No.6, pp.479–86, accessed at www.ncbi.nlm.nih.gov/pmc/articles/PMC4918728, 21 January 2019.

9 www.washingtonpost.com/lifestyle/food/what-is-dude-food-anyway-we-asked-the-experts-and-they-fired-away/2015/06/19/91da5fdc-15b5-11e5-89f3-61410da94eb1_story.html.

10 Billing, J., Sherman, P.W., 'Antimicrobial functions of spices: Why some like it hot', *The Quarterly Review of Biology*, 1998, Vol.73, No.1.

11 Mujcic, R. et al., 'Evolution of well-being and happiness after increases in consumption of fruit and vegetables', *Am J Public Health*, 2016, Vol.106, No.8, pp.1504–10, accessed at www.ncbi.nlm.nih. gov/pmc/articles/PMC4940663, 21 January 2019.

12 www.abc.net.au/news/2016-08-03/touch-screens-impacting-on-kids-writing-skills-therapist-says/7683054.

13 Keifer, M., Schuler, S, Mayer, C. Trumpp, N.M., Hille, K., Sachse, S., 'Handwriting or typewriting? The influence of pen- or keyboard-based writing training on reading and writing performance in preschool children', *Adv Cogn Psychol.*, 2015, Vol.11, No.4, pp.136–46.

14 Jakubiak, B.K., 'Affectionate touch to promote relational, psychological and physical wellbeing in adulthood: A theoretical model and review of the research', *Personality and Social Psychology Review*, 2016, Vol.21, No.3, pp.228–52. accessed at journals.sagepub. com/doi/10.1177/1088868316650307.

15 Degges-White, S., 'Skin hunger: Why you need to feed your hunger for contact', *Psychology Today*, 2015, www.psychologytoday.com/ blog/lifetime-connections/201501/skin-hunger-why-you-need-feed-your-hunger-contact.

16 utnews.utoledo.edu/index.php/08_08_2018/ut-chemists-discover-how-blue-light-speeds-blindness.

Epilogue

1 Wallisch, P., 'Illumination assumptions account for individual differences in the perceptual interpretation of a profoundly ambiguous stimulus in the color domain: "The dress"', *Journal of Vision*, 2017, Vol.17, No.4.

2 Wallisch, 2017.

Postscript

1 Parma, V., et al., 'More than Smell: COVID-19 is Associated with Severe Impairment of Smell, Taste and Chemesthesis', *Chemical Senses*, 2020, Vol. 45, No. 7, pp.609–22.

2 Speth, M., et al., 'Mood, Anxiety and Olfactory Disorder in COVID-19: Evidence of Central Nervous System Involvement?', *The Laryngoscope*, 2020, Vol. 130, pp.2520–5.

3 Koumpa, F.S., et al., 'Sudden Irreversible Hearing Loss Post COVID-19', *BMJ Case Reports CP*, 2020, 13, e238419.

4 Chen, L., Liu, M., Zhang, Z., et al., 'Ocular Manifestations of a Hospitalised Patient with Confirmed 2019 Novel Coronavirus Disease', *British Journal of Ophthalmology*, 2020, 104, pp.748–51.

5 Carel, H., et al., 'Reflecting on Experiences of Social Distancing', *The Lancet*, 2020, Vol. 396, Issue 10244, pp.87–8, www.thelancet.com/journals/lancet/article/PIIS0140-6736(20)31485-9/fulltext.

6 *Independent*, 18 September, 2020, www.independent.co.uk/news/world/americas/us-politics/trump-coronavirus-good-thing-shake-hands-disgusting-people-olivia-troye-b480406.html, accessed 19 September 2020.

7 Cohen, S., et al., 'Does Hugging Provide Stress-buffering Social Support? A Study of Susceptibility to Upper Respiratory Infection and Illness', *Psychol Sci.*, 2015, Vol. 26(2), pp.135–47.

8 www.iea.org/reports/data-centres-and-data-transmission-networks, accessed 19 November 2020.

9 www.ofcom.org.uk/about-ofcom/latest/features-and-news/uk-internet-use-surges, accessed 19 November 2020.

10 Wai, A., Wong, C., et al., 'Digital Screen Time During COVID-19 Pandemic: Risk for a Further Myopia Boom?', *American Journal of Ophthalmology*, 2020, doi.org/10.1016/j.ajo.2020.07.034.

BIBLIOGRAPHY

Adkins, L.A., *Empires of the Plain*, citing Borger, R. (1975–78) from the RAS archive (London: Harper Perennial, 2004).

Alexander [ed.], *The Painted Page: Italian Renaissance Book Illumination 1450–1550* (London: Prestel, 1994).

Asimov, I., *Eyes on the Universe: A History of the Telescope* (London: Andre Deutch, 1976).

Bacon, R., *Opus Major*, Introduction and trans. Bridges, J.H. (London: Williams and Norgate, 1900 [1267]).

Baragli, S., *European Art of the Fourteenth Century* (Los Angeles: Getty Publications, 2007).

Baxandall, M., *Painting and Experience in Renaissance Italy* (Oxford: Oxford University Press, 1972).

Benjamin, W., *The Work of Art in the Age of Mechanical Reproduction* (London: Penguin Great Ideas, 1936).

Berger, J., *Ways of Seeing* (London: Bloomsbury, 1972).

Berger, J., *About Looking* (London: Bloomsbury, 1980).

Bewernick, H., *The Storyteller's Memory Palace: A Method of Interpretation Based on the Function of Memory Systems in Literature* (Frankfurt am Maim: Peter Lang, 2010).

Blackwell, R.J., *Behind the Scenes at Galileo's Trial* (Notre Dame, IN: University of Notre Dame Press, 2006).

Bloom, J., *Paper Before Print: The History and Impact of Paper in the Islamic World* (New Haven, CT: Yale University Press, 2001).

Blunt, W., Raphael, S., *The Illustrated Herbal* (London: Frances Lincoln Publishers, 1979).

Bowser, E., *The Transformation of Cinema* (Berkeley, CA: University of California Press, 1994).

Brox, Jane, *Brilliant: The Evolution of Artificial Light* (London: Souvenir Press, 2010).

Bruce, V., et al., *Visual Perception: Physiology, Psychology and Ecology* (Hove and New York: Psychology Press, 2003).

Brunel, Elliette, Chauvet, Jean-Marie, Hillaire, Christian, *The Discovery of the Chauvet-Pont d'Arc Cave* (Saint-Remy-de-Provence: Editions Equinoxe, 2014).

Buckland, G., *Fox Talbot and the Invention of Photography* (London: Scholar Press, 1980).

Burns, E., *Invasion of the Mind Snatchers* (Philadelphia, PA: Temple University Press, 2010).

Burton, F.D., *Fire: The Spark that Ignited Human Evolution* (Albuquerque, NM: UNMP, 2009).

Camille, M., *Mirror in Parchment: The Luttrell Psalter and the Making of Medieval England* (Chicago, IL: University of Chicago Press, 1997).

Cantor, N.F., *The Civilisation of the Middle Ages* (New York, NY: Harper, 1994).

Capra, F., *The Name above the Title* (New York: Bantam Books, 1971).

Carpenter, E.S., *Oh, What a Blow That Phantom Gave Me!* (New York, NY: Holt, Rinehart and Winston, 1972).

Chapman, A., *Stargazers: Copernicus, Galileo, the Telescope and the Church* (Oxford: Lion Hudson, 2014).

Chauvet, J.-M., Brunel Deschamps, E., Hillaire, C., *Chauvet Cave: The Discovery of the World's Oldest Paintings* (London: Thames and Hudson, 1996).

Clark, G., *A Farewell to Alms: A Brief Economic History of the World* (Princeton, NJ: Princeton University Press, 2008).

Clark, S., *Vanities of the Eye: Vision in Early Modern European Culture* (Oxford: Oxford University Press, 2007)

Clottes, J., *Cave Art* (London: Phaidon, 2008).

Cooley, C.H., *Human Nature and the Social Order* (New York, NY: Charles Scribner's Books, 1902).

Darwin, C., *On the Origin of Species* (First Edition) (London: Murray, 1859).

Darwin, C., *The Descent of Man, and Selection in Relation to Sex*, Vol. 1 (London: John Murray, 1871).

Defoe, D., *History of the Devil* (Boston, MA: C.D. Strong, 1848 [1726]).

de Font-Reaulx, D., *Painting and Photography: 1839–1914* (Paris: Flammarion, 2012).

de Santillana, G., *The Crime of Galileo* (Chicago, IL: University of Chicago Press, 1955).

Douglass, F., *Narrative of the Life of Frederick Douglass* (Amazon edition, 2013) (Boston: Anti-Slavery Office: 1845).

Duffy, B., *The Perils of Perception* (London: Atlantic Books, 2018).

Dunbar, R., *Human Evolution* (London: Pelican, 2014).

Dunbar, R.I.M., Gowlett, J.A.J., *From Lucy to Language: The Benchmark Papers*, eds Dunbar, Gamble, Gowlett (Oxford: Oxford University Press, 2014).

Dunn, R., *The Telescope: A Short History* (London: National Maritime Museum, 2009).

Eamon, W., *Science and the Secrets of Nature: Books of Secrets in Medieval and Early Modern Culture* (Princeton, NJ: Princeton University Press, 1994).

Eco, U., *Art and Beauty in the Middle Ages* (New Haven, CT: Yale University Press, 1986).

Einhard, *The Life of Charlemagne*, trans. Turner, S.E. (New York, NY: Harper & Brothers, 1880) accessed at Fordham University Medieval Sourcebook, sourcebooks.fordham.edu/basis/einhard. asp#Charlemagne%20Crowned%20Emperor.

Eisenstein, E.L., *The Printing Revolution in Early Modern Europe* (Cambridge: Cambridge University Press, 1983).

Ekirch, A.R., *At Day's Close: A History of Nighttime* (London: Phoenix, 2005).

Erickson, C., *The Medieval Vision: Essays in History and Perception* (New York, NY: Oxford University Press, 1976).

Foer, Joshua, *Moonwalking with Einstein: The Art and Science of Remembering Everything* (London: Penguin Press, 2011).

Fox Talbot, H., *The Pencil of Nature* (London: Longman, Brown, Green and Longmans, 1844), accessed via Gutenburg.org at www. gutenberg.org/files/33447/33447-pdf.pdf, 20 July 2018.

Frazer, J.G., *Myths of the Origin of Fire: An Essay* (London: Macmillan and Co. Limited, 1930).

Frisby, J.P., *Seeing: Illusion, Brain and Mind* (Oxford: Oxford University Press, 1979).

Frugoni, C., *Inventions of the Middle Ages*, trans. McCuaig, W. (London: The Folio Society, 2007).

Galassi, P., *Before Photography: Painting and the Invention of Photography* (New York, NY: Museum of Modern Art, 1981)

Gombrich, E.H., *The Story of Art* (London: Phaidon, 1950).

Gombrich, E.H., *The Story of Art* (Fifteenth Edition) (London: Phaidon, 1989).

Gombrich, E.H., *Art and Illusion* (London: Phaidon, 2002).

Greenblatt, S., *The Swerve: How the World Became Modern* (London: Norton Books, 2001).

Gregory, R.L., *Eye and Brain: The Psychology of Seeing* (Oxford: Oxford University Press, 1998).

Griffiths, J., *The Third Man: The Life and Times of William Murdoch, Inventor of Gaslight* (London: Andre Deutsch, 1992).

Guthrie, R.D., *The Nature of Paleolithic Art* (Chicago, IL: Chicago University Press, 2006).

Hall, J., *The Self Portrait: A Cultural History* (London: Thames and Hudson, 2014).

Herbert, K., *Looking for the Lost Gods of England* (Cambridgeshire: Anglo Saxon Books, 1994).

Hills, P., *The Light of Early Italian Painting* (New Haven, CT and London: Yale University Press, 1987).

Hockney, D., *Secret Knowledge: Rediscovering the Lost Techniques of the Old Masters* (London: Thames and Hudson, 2006).

Hooke, R., *Micrographia, or, Some Physiological Descriptions of Minute Bodies Made with Magnifying Glasses, With Observations and Inquiries Made Thereon* (London: Royal Society, 1665), accessed via archive. org/stream/mobot31753000817897#page/210/mode/2up.

Hoskin, M. (ed.), *The Cambridge Concise History of Astronomy* (Cambridge: Cambridge University Press, 1997).

Hughes, B., *Istanbul: A Tales of Three Cities* (London: Bloomsbury, 2017).

Ilardi, V., *Renaissance Vision: From Spectacles to Telescopes* (Philadelphia, PA: American Philosophical Society, 2007).

Ings, S., *The Eye: A Natural History* (London: Bloomsbury Publishing, 2007).

Ivins, W.M., *Prints and Visual Communication* (Cambridge, MA: MIT Press, 1953).

Jones, C., *Paris: Biography of a City* (London: Penguin Books, 2006).

Kuhn, A., *An Everyday Magic: Cinema and Cultural Memory* (London and New York, NY: IB Taurus, 2002).

Kuhn, R., *The Media in France* (London and New York, NY: Routledge, 1995).

Land, M., *The Eye: A Very Short Introduction* (Oxford: Oxford University Press, 2014).

Layard, R., *Happiness* (London: Penguin Books, 2005).

Lefèvre, W. (ed.), *Inside the Camera Obscura: Optics and Art under the Spell of the Projected Image* (Berlin: Max Planck Institute for the History of Science, 2007).

Lehr, D., *The Birth of a Nation* (New York, NY: Public Affairs, 2014).

Lester, L.K. (ed.), *Plague and the End of Antiquity* (Cambridge: Cambridge University Press, 2007).

Leverington, D., *Babylon to Voyager and Beyond: A History of Planetary Astronomy* (Cambridge: Cambridge University Press, 2003).

Lindberg, D., *Theories of Vision from Al Kindi to Kepler* (Chicago, IL: University of Chicago Press, 1976).

Lindberg, D.C., Westman, R.S., *Reappraisals of the Scientific Revolution* (Cambridge: Cambridge University Press, 1990).

Litchfield, R.B., *Tom Wedgwood, the First Photographer: An Account of his Life, his Discovery and his Friendship with Samuel Taylor Coleridge, Including the Letters of Coleridge to the Wedgwoods and an Examination of Accounts of Alleged Earlier Photographic Discoveries* (London: Duckworth and Co., 1903), p.24, accessed via archive.org/stream/tomwedgwood first00litcrich#page, 20 July 2018.

Livingstone, M., *Vision and Art: The Biology of Seeing* (Second Edition) (New York, NY: Abrams, 2014).

Lochie, K., *Covert Operations: The Medieval Uses of Secrecy* (Philadelphia, PA: University of Pennsylvania Press, 1999).

Maas, J., *Victorian Painters* (London: Barrie and Jenkins, 1978).

Magno, A.M., *Bound in Venice: The Serene Republic and the Dawn of the Book* (New York, NY: Europa Editions, 2013).

McCarthy, A., *The Citizen Machine: Governing by Television in 1950s America* (New York, NY: The New Press, 2010).

McLuhan, M., *The Gutenberg Galaxy: The Making of Typographical Man* (Toronto: University of Toronto Press, 1962).

Mellaart, J., *Catal Hoyuk: A Neolithic Town in Anatolia* (New York, NY: McGraw Hill, 1967).

Meltzner, M., *The Printing Press* (Salt Lake City, UT: Benchmark Books, 2004) (accessed via www.questia.com/read/122746489/the-printing-press).

Meredith, R., *Mr. Lincoln's Camera Man: Mathew B. Brady* (New York, NY: Dover, 1974).

Minnaert, M., *The Nature of Light and Colour in the Open Air* (New York, NY: Dover Publications, 1954).

Mirzoeff, N., *How to See the World* (London: Penguin Random House, 2015).

Moran, J., *Armchair Nation* (London: Profile Books, 2013).

Ong, W., *Orality and Literacy: The Technologizing of the Word* (Thirtieth Anniversary Edition) (London and New York, NY: Routledge, 1982).

Parker, A., *In the Blink of an Eye: How Vision Sparked the Big Bang of Evolution* (New York, NY: Basic Books, 2003).

Pendergrast, M., *Mirror Mirror: A History of the Human Love Affair with Reflection* (New York, NY: Basic Books, 2003).

Pesic, P., *Sky in a Bottle* (Cambridge, MA: MIT Press, 2007).

Portman, N., *Amusing Ourselves to Death: Public Discourse in the Age of Show Business* (New York, NY: Viking Press, 1985).

Postman, N., *Technopoly: The Surrender of Culture to Technology* (New York, NY: Knopf, 1993).

Powell, B., *Writing: Theory and History of the Technology of Civilisation* (Hoboken, NJ: Wiley, 2012).

Quinn, J., *In Search of the Phoenicians* (Oxford: Oxford University Press, 2017).

Reeves, E., *Galileo's Glassworks: The Telescope and the Mirror* (Cambridge, MA: Harvard University Press, 2008), accessed via www.questia.com/read/119165310/galileo-s-glassworks-the-telescope-and-the-mirror, 15 April 2018.

Robinson, A., *The Last Man Who Knew Everything: Thomas Young* (Oxford: Oneworld, 2006).

Rouillé, A., Lemagny, J.-C., *A History of Photography* (Cambridge: Cambridge University Press, 1988).

Rublack, U., *The Astronomer and the Witch* (Oxford: Oxford University Press, 2015).

Sacks, O., *The Mind's Eye* (New York, NY: Knopf, 2010).

Saengar, P., *Space Between Words: The Origin of Silent Reading* (Stanford, CA: Stanford University Press, 1997).

Schiller, G., *Iconography of Christian Art, Vol. II* (English translation from German) (London: Lund Humphries, 1972).

Schivelbusch, W., *Disenchanted Night: The Industrialization of Light in the Nineteenth Century* (Berkeley, CA: University of California Press, 1995).

Shaviro, S., *The Cinematic Body* (Minneapolis, MN: University of Minnesota Press, 1993).

Shipman, P., *The Invaders: How Humans and Their Dogs Drove Neanderthals to Extinction* (Boston, MA: Harvard University Press, 2005).

Sinisgalli, Rocco, *Perspective in the Visual Culture of Classical Antiquity* (Cambridge: Cambridge University Press, 2012).

Sobel, D., *A More Perfect Heaven: How Copernicus Revolutionized the Cosmos* (London: Walker & Company, 2012).

Stauffer, J., Trodd, Z., Bernier, C-M., *Picturing Frederick Douglass* (Revised Edition) (New York, NY: W.W. Norton and Company Inc., 2015).

Steadman, P., *Vermeer's Camera* (Oxford: Oxford University Press, 2001).

Stephens, M., *The Rise of the Image and the Fall of the Word* (Oxford: Oxford University Press, 1998).

Taylor, J.E., *Christians and the Holy Places* (Oxford: Oxford University Press, 1993).

Temple, R., *Crystal Sun* (London: Century, 2000).

Thomson, J., *The Scot Who Lit the World: The Story of William Murdoch, Inventor of Gas Lighting* (self-published, 2003).

Toohey, P., *Melancholy, Love and Time* (Ann Arbor, MI: University of Michigan Press, 2004).

Twenge, J.M., *The Narcissism Epidemic: Living in the Age of Entitlement* (New York: Atria Books, 2010).

Twenge, J.M., *Generation Me: Why Today's Young Americans Are More Confident, Assertive, Entitled – And More Miserable Than Ever Before* (New York, NY: Atria Books, updated 2014).

Uhlendorf, A., *The Invention of Printing and its Spread Until 1470* (Michigan: Michigan University Press, 1932).

Van Huyssteen, J.W., Wiebe, E.P. (eds), *In Search of Self: Interdisciplinary Perspectives on Personhood* (Cambridge: Eerdmans Publishing, 2011).

Walsh, J.J., *The World's Debt to the Irish* (Mounet Sud, France: Tradibooks, 2010).

Welch, D., *Propaganda and the German Cinema, 1933–1945* (London and New York, NY: IB Taurus, 1983).

Wells, H.G., *When the Sleeper Wakes* (London: Collins, 1910).

Whitehouse, D., *Glass: A Short History* (Washington DC: Smithsonian Institutions, 2012).

Williams, A., *Republic of Images* (Cambridge, MA: Harvard University Press, 1992).

Wrangham, R., *Catching Fire: How Cooking Made Us Human* (New York, NY: Basic Books, 2009).

Yates, F.A., *The Art of Memory* (London: Routledge, 1999).

Young, S., *AD 500* (London: Wiedenfeld and Nicholson, 2005).

Zajonc, A., *Catching the Light: What is Light and How Do We See It?* (Oxford: Oxford University Press, 1993).

Zelanski, P., Fisher, M.P., *The Art of Seeing* (Englewood Cliffs, NJ: Prentice-Hall, 1988).

Zuboff, S., *The Age of Surveillance Capitalism* (New York, NY: Public Affairs, 2019).

Other References (Articles, Press, Websites, Videos, etc.)

Aiello, L., Wheeler, P., 'The expensive-tissue hypothesis: the brain and the digestive system in human and primate evolution', *Current Anthropology*, 1995, Vol.36, pp.199–221.

Al Kahlili, J., BBC News, 2009, accessed at news.bbc.co.uk/1/hi/sci/tech/7810846.stm.

Anadolu Agency, 'Çatalhöyük excavations reveal gender equality in ancient settled life', 2014, hurriyetdailynews.com, citing interview with Ian Hodder.

Andrade, E.N. da C., 'Galileo', *Notes and Records of the Royal Society of London*, 1964, Vol.19, No.2, pp.120–30, accessed at www.jstor.org/stable/3519848, 20 May 2018.

Anon, *Eye-Witness Account of Image-Breaking at Antwerp, 21–23 August 1566*, University of Leiden, accessed at dutchrevolt.leiden.edu/english/sources/Pages/15660721.aspx.

Art [and History] by Lightning Flash: The Birth of a Nation and Black Protest, Roy Rosenzweig Center for History and New Media, George Mason University, accessed at chnm.gmu.edu/episodes/the-birth-of-a-nation-and-black-protest/, 7 January 2019.

Bacon, R., *Opus Majus*, 1267, cited in Reeves, E., 2008.

Banks, M.S., Sprague, E.W., Schmoll, J., Parnell, J.A.Q. Love, G.D., 'Why do animal eyes have pupils of different shapes?', *Science Advances*, 2015, Vol. 1, No. 7.

Barinaga, M., 'Focusing on the eyeless gene', *Science*, 1995, Vol.267, No. 5205, pp. 1766–67.

BBC Radio, 'The Carolingian Renaissance', *In Our Time*, 2006, accessed at www.bbc.co.uk/programmes/p003hydz.

Beaman, A.L., Klentz, B., Diener, E., Svanum, S., 'Self-awareness and transgression in children', *Journal of Personality and Social Psychology*, 1979, Vol. 37, No. 10, pp.1835–46.

Benbow, M., 'Birth of a quotation: Woodrow Wilson and "like writing history with lightning"', *The Journal of the Gilded Age and Progressive Era*, 2010, Vol.9, No.4, pp.509–33.

Bianchi, R.S., 'Reflections in the sky's eyes', *Notes on the History of Science*, 2005, Vol. 4.

Billau, C., 'UT Chemists discover how blue light speeds blindness', *University of Toledo News*, 8 August 2018, utnews.utoledo.edu/index.php/08_08_2018/ut-chemists-discover-how-blue-light-speeds-blindness.

Billing, J., Sherman, P.W., 'Antimicrobial functions of spices: why some like it hot', *The Quarterly Review of Biology*, 1998, Vol.73, No.1.

Birth of a Nation film viewed at ia902702.us.archive.org/29/items/dw_griffith_birth_of_a_nation/birth_of_a_nation_512kb.mp4

Birth of a Nation film poster accessed at commons.wikimedia.org/w/index.php?search=birth+of+a+nation&title=Special:Search&go=Go&searchToken=qrkxlrnmpsmpozhhwximo2bg#/media/File:Birth_of_a_Nation_-_Academy.jpg.

Block, R., *Live from Macworld*, 2007, www.engadget.com/2007/01/09/live-from-macworld-2007-steve-jobs-keynote, accessed at bgr.com/2016/07/01/iphone-reviews-original-negative-ballmer-dvorak; www.engadget.com/2007/07/25/apple-sold-270-000-iphones-in-the-first-30-hours, 18 January 2019.

Blood Lake: Attack of the Killer Lampreys, www.imdb.com/title/tt3723790.

Bonser, W., 'The cult of relics in the Middle Ages', *Folkore*, 1962, Vol.73, No.4, pp.234–56.

Boring, E.G., Dual role of the Zeitgeist in scientific creativity, *Scientific Monthly*, 1950, Vol.80, pp.101–06.

Boucicault, D., *The Octoroon*, 1859, accessed at archive.org/stream/octoroonorlifein00bouc/octoroonorlifein00bouc_djvu.txt referenced in www.phrases.org.uk/meanings/camera-cannot-lie.html, accessed 10 December 2018. www.phrases.org.uk/meanings/camera-cannot-lie.html, accessed 10 December 2018. en.wikipedia.org/wiki/A_picture_is_worth_a_thousand_words#cite_note-1.

Bower, J., '5 rebuttals to 'that' smartphone article in *The Atlantic*', 2017, www.digitalnutrition.com.au/5-rebuttals-to-that-smartphone-article-in-the-atlantic.

Bowmaker, J.K., 'Evolution of colour vision in vertebrates', *Eye*, 1998, Vol. 12 (3b), pp. 541–47.

Boyd, D.M., Ellison, N.B., 'Social network sites: definition, history, and scholarship', 2007, *Journal of Computer-Mediated Communication*,

Vol.13, No.1, accessed at onlinelibrary.wiley.com/doi/full/10.1111/
j.1083-6101.2007.00393.x?scrollTo=references, 18 January 2019.
mashable.com/video/blood-wolf-moon.

British Astronomical Association's Campaign for Dark Skies at www.
britastro.org/dark-skies and in the USA the International Dark-Sky
Association at www.darksky.orgm the US Library of Congress,
www.loc.gov/item/mfd.22004

Bruner, E., Lozano, M., 'Extended mind and visua-spatial integration:
three hands for the Neanderthal lineage', *Journal of Anthropological
Science*, 2014, Vol. 92, pp.273–80.

Butsch, W., 'The making of American audiences', *International Labor and
Working-Class History*, 2000, No.59, Workers and Film: As Subject
and Audience (Cambridge: Cambridge University Press, 2001).

Calmo, C., 'The big question: what is the Rosetta Stone and
should Britain return it to Egypt?', *The Independent*, 2009, www.
independent.co.uk/news/uk/this-britain/the-big-question-what-is-
the-rosetta-stone-and-should-britain-return-it-to-egypt-1836610.
html, accessed 21 October 2018.

Campbell, et al., *Tuning in, tuning out revisited: a closer look at the causal
links between television and social capital*, 1999, citing studies from the
US, UK, Australia and South Africa, accessed at sites.hks.harvard.
edu/fs/pnorris/Acrobat/TVAPSA99.PDF, 15 January 2019.

Carothers, J.C., 'Psychiatry and the written word', *Psychiatry*, 1959,
pp.304–18.

Carpenter, E.S., 'The tribal terror of self-awareness', in Hockings, P. (ed.),
Principles of Visual Anthropology (The Hague: Mouton, 1975), p.455.

Carpino, Alexandra A., 'Reflections from the tomb: mirrors as grave
goods in late classical and Hellenistic Tarquinia', *Etruscan Studies*,
2008, Vol.11, No.1, accessed at scholarworks.umass.edu/etruscan_
studies/vol11/iss1/1.

Cashdan, E.A., 'Egalitarianism among hunters and gatherers',
American Anthropologist, 1980, Vol.82, No.1, pp.116–20, accessed at
onlinelibrary.wiley.com/doi/10.1525/aa.1980.82.1.02a00100/full,
1 February 2016.

Chadwick, Richard, 'Who were the skywatchers of the ancient near
east?', *Bulletin of the Canadian Society for Mesopotamian Studies*, 1988,
No.39, pp.5–14.

Chandler, J., 'Little common ground as land grab splits people', *Sydney
Morning Herald*, 15 October 2011, accessed at www.smh.com.
au/world/little-common-ground-as-land-grab-splits-a-people-
20111014-1lp09.html, 1 February 2016.

Chaochao, G., et al., 'Reconciling multiple ice-core volcanic histories: the potential of tree-ring and documentary evidence, 670–730 CE', *Quaternary International*, 2016, Vol.394, pp.180–93.

Charles Falco's website at wp.optics.arizona.edu/falco/art-optics/ for information and links on the Hockney–Falco thesis.

Chau, H.F., Boland, J.E., Nisbett, R.E., 'Cultural variation in eye movements during perception', *Proceedings of the National Academy of Sciences of the United States of America*, 2005, Vol. 102, No. 35, pp.12629–633.

Clark, J.D., Harris, J.W.K., 'Fire and its roles in early hominid lifeways', *African Archaeological Review*, Vol. 3, No.1, pp.3–27 (Springer, 1985).

Cook, D.A., Sklar, R. 'History of the motion picture', *Encyclopaedia Britannica*, accessed at www.britannica.com/art/history-of-the-motion-picture/The-silent-years-1910-27, 12 January 2019.

Copernicus, N., *Commentariolus*, 1415 (trans. Rosen), accessed at copernicus.torun.pl/en/archives/astronomical/1/?view=transkryp cja&lang=en, accessed via themcclungs.net/physics/download/H/Astronomy/Commentariolus%20Text.pdf, 16 May 2018.

Copernicus, N., *de Revolutionibus*, 1543, Preface by Andreas Osiander, accessed at www.geo.utexas.edu/courses/302d/Fall_2011/Full%20text%20-%20Nicholas%20Copernicus,%20_De%20Revolutionibus%20%28On%20the%20Revolutions%29,_%201.pdf, 30 May 2018.

Cromby, I., 'The madonna of the future', *Art Journal*, 2004, Vol.43, National Gallery of Victoria, accessed at Carletonwatkins.org.

Curtin. S.C. et al., 'Increase in suicide in the United States, 1999–2014', *NCHS Data Brief*, 2016, No. 241: 1–8, accessed at www.ncbi.nlm.nih.gov/pubmed/27111185, 20 January 2019. www.pewinternet.org/2018/11/28/teens-and-their-experiences-on-social-media, www.marketwatch.com/story/want-to-delete-facebook-read-what-happened-to-these-people-first-2018-07-27.

Damerow, P., 'The origins of writing as a problem of historical epistemology', *Cuneiform Digital Library Journal*, 2006:1 cdli.ucla.edu/pubs/cdlj/2006/cdlj2006_001.html © Cuneiform Digital Library Initiative.

Davy, H., 'An account of a method of copying paintings upon glass, and of making profiles, by the agency of light upon nitrate of silver, with observations by Humphrey Davy. Invented by T. Wedgwood, Esq.', *Journals of the Royal Institution*, 1802, Vol.1.

De Beaune, S., 'Palaeolithic lamps and their specialization: a hypothesis', *Current Anthropology*, 1987, Vol.28, No.4, pp.569–77. Accessed at 0-www.jstor.org.wam.city.ac.uk/stable/2743501.

De Beaune, S., White, R., 'Ice Age lamps', *Scientific American*, 1993, March issue, accessed at www.academia.edu/416951/Ice_Age_Lamps.

Degges-White, S., 'Skin hunger: why you need to feed your hunger for contact', *Psychology Today*, 2015, www.psychologytoday.com/blog/lifetime-connections/201501/skin-hunger-why-you-need-feed-your-hunger-contact.

Deleuze, G., *Cinéma I: L'image-mouvement*, 1983, Trans. (1986) *Cinema 1: The Movement-Image* (1986).

Descartes, R., *Discourse of the Method of Rightly Conducting the Reason, and Seeking Truth in the Sciences*, 1637, accessed at www.gutenberg.org/files/59/59-h/59-h.htm.

Diener, E., Wallbom, M., 'Effects of self awareness on antinormative behaviour', *Journal of Research in Personality*, 1976, Vol.10, pp.107–11.

Druckman, J.N., 'The power of television images: the first Kennedy-Nixon debate revisited', *The Journal of Politics*, 2003, Vol.65, No.2, pp.559–71.

Dunbar, R.I.M., 'The social brain: mind, language, and society in evolutionary perspective', *Annual Review of Anthropology*, 2003, Vol.32, pp.163–81, www.jstor.org.wam.city.ac.uk/stable/25064825.

Dunbar, R.I.M., 'The social brain hypothesis and its implications for social evolution', *Annals of Human Biology*, 2009, Vol.36, No.5, pp.562–72.

D'Ooge, C., *Symposium on 'Birth of a Nation'*, US Library of Congress, 1994, accessed at www.loc.gov/loc/lcib/94/9413/nation.html

Early Television Museum, accessed at www.earlytelevision.org/prewar.html.

Edgerton, S.Y., 'Brunelleschi's mirror, Alberti's window, and Galileo's "perspective tube"', *História Ciências Saúde-Manguinhos*, 2006, Vol.13.

Egan, G., *Hearing or Seeing a Play?: Evidence of Early Modern Theatrical Terminology*, 2001, accessed at gabrielegan.com/publications/Egan2001k.htm, 27 March 2018.

Elrasheedy, A, & Schindler, D., 'Illuminating the past: exploring the function of ancient lamps', *Near Eastern Archaeology*, 2015, Vol.78, No.1, pp.3–42.

Englund, R.K., 'The Ur III collection of the CMAA', *Cuneiform Digital Library Journal*, 2002:1 cdli.ucla.edu/pubs/cdlj/2002/001.html © Cuneiform Digital Library Initiative.

Facebook investor relations statistics, accessed at zephoria.com/top-15-valuable-facebook-statistics, 18 January 2019; accessed at www.

rawhide.org/blog/infographics/selfie-obsession-the-rise-of-social-media-narcissism, 20 January 2019.

Falco, C., 'Optics and renaissance art', in *Optics in Our Time*, 2006.

Fiorentini, E., *Camera Obscura vs. Camera Lucida: Distinguishing Nineteenth Century Modes of Seeing*, Max Plank Institute for History of Science, 2006, Vol.307, p.7, accessed online at www.researchgate.net/publication/41125012_Camera_obscura_vs_camera_lucida_-_distinguishing_early_nineteenth_century_modes_of_seeing, 11 July 2018.

Ford, B.J., 'Scientific illustration in the eighteenth century', Chapter 24 in Porter, R. (ed.), *The Cambridge History of Science* (Cambridge: Cambridge University Press, 2003).

Fossil gallery, Burgess Shale website, Royal Ontario Museum, burgess-shale.rom.on.ca/en/fossil-gallery/index.php.

Freud Museum, www.freud.org.uk/2018/07/23/self-reflection-mirrors-in-sigmund-freuds-collection.

Fujino, S. et al., 'Job Stress and Mental Health Among Permanent Night Workers', *Journal of Occupational Health*, 2001, Vol.43, pp.301–06.

Galileo, G. (trans. Drake, S.), *The Assayer*, 1623, accessed at web.stanford.edu/~jsabol/certainty/readings/Galileo-Assayer.pdf.

Galileo, G., 'Letter to the Grand Duchess Christina of Tuscany', 1615, accessed at sourcebooks.fordham.edu/mod/galileo-tuscany.asp, 12 June 2018.

Galileo, G., *The Starry Messenger*, 1610, The Galileo Project>Chronology>Timeline, accessed at galileo.rice.edu/chron/galileo.html

Gallese, V., Guerra, M., 'Embodying movies: embodied simulation and film studies', *Cinema*, 2012, Vol.3, pp.183–210, accessed at core.ac.uk/download/pdf/96698686.pdf, 7 January 2019.

Gallup, G.G. Jr, 'Self awareness and emergence of mind in primate', *American Journal of Primatology*, 1982, Vol.2, pp.237–48.

Gani, M.R., Gani N.D.S., *Geotimes*, 2008, Vol. 1, accessed at www.geotimes.org/jan08/article.html?id=feature_evolution.html, May 2016.

Geffney, V. et al., 'Time and a place: a luni-solar "time-reckoner" from 8th millennium BC Scotland', *Internet Archaeology*, 2013, Vol.34, accessed at intarch.ac.uk/journal/issue34/gaffney_index.html, 14 May 2018.

Gehring, W., Ikeo, K., 'Pax6: Mastering eye morphogenesis and eye evolution', *Trends in Genetics*, 1999, Vol. 15, No. 9.

George, A.R., 'The gods Išum and Ḫendursanga: night watchmen and street-lighting in Babylonia', *Journal of Near Eastern Studies*, 2015, Vol.74, No.1.

Gest, H., 'The discovery of microorganisms by Robert Hooke and Antoni van Leeuwenhoek, fellows of The Royal Society', *Notes Rec. R. Soc.* London, 2004, Vol.58 pp.187–201.

Gingerich, O., 'Did Copernicus Owe a Debt to Aristarchus?', *Journal for the History of Astronomy*, 1985, Vol. 16, No. 1.

Goldfarb, S.J., 'A Regency Gas Burner', *Technology sand Culture*, 1971, Vol.12, No.3, p.476.

Goodman, K., McCravy, K.W., 'Pyrophilious insects', *Encyclopaedia of Entomology*, ed. Capinera, J.L. Ed., Springer (The Netherlands: Springer, 2008), pp.3090–93.

Gordon, T. (trans. from the original), *Tacitus on Germany*, 1910, via Project Gutenberg, accessed at www.gutenberg.org/files/2995/2995. txt, 5 November 2017.

Gossman, L., *The Important Influence of Paintings in Early Photography*, 2015, referencing his book *Thomas Annan of Glasgow: Pioneer of the Documentary Photograph*, accessed at brewminate.com/the-important-influence-of-paintings-in-early-photography.

Gowlett, J.A.J., Wrangham, R.W., 'Earliest fire in Africa: towards the convergence of archaeological evidence and the cooking hypothesis', *Azania: Archaeological Research in Africa*, 2013, Vol.48, No.1, pp.5–30.

Greilsammer, Myriam, 'The midwife, the priest, and the physician: the subjugation of midwives in the low countries at the end of the Middle Ages', *The Journal of Medieval and Renaissance Studies*, 1991, Vol.21, pp.285–329.

Guild of Scholars website, gofs.co.uk/dynamicpage.aspx?id=84.

Hagopian, K.J. (undated), *Film notes*, New York State Writers' Institute, www.albany.edu/writers-inst/webpages4/filmnotes/fns07n6.html, and Pierpont, C.R., 2015, review of Wieland, K., 'Dietrich & Riefenstahl: Hollywood, Berlin, and a Century in Two Lives', 2015, Liveright, accessed at www.newyorker.com/magazine/2015/10/19/bombshells-a-critic-at-large-pierpont.

Halder, G., Callaerts, P., Gehring, W.J., 'Induction of ectopic eyes by targeted expression of the eyeless gene in Drosophila', *Science*, 1995, Vol. 267, No. 5205, pp.1788–92.

Harford, T., *50 Things That Made the Modern Economy*, BBC World Service, 2017, accessed at www.bbc.co.uk/news/business-41582244, 2 April 18.

Harkness, D.E., 'Alchemy and eschatology: exploring the connections between John Dee and Sir Isaac Newton in force', Popkin, R.H. (ed.) *Newton and Religion, Context Nature and Influence* (Berlin: Springer-Science and Business Media, 1999).

Hasson, U. et al., 'Neurocinematics: the neuroscience of film', *Projections*, 2009, Vol.2, No.1, Summer 2008: 1–26, Berghahn Journals.

Haynes, T., *Dopamine, Smartphones and You: A Battle for your Time*, 2018, sitn.hms.harvard.edu/flash/2018/dopamine-smartphones-battle-time.

Heath, Sir T. (trans.), Aristarchus of Samos, 1913, Oxford, p.302. Cited in Gingerich, O., 'Did Copernicus owe a debt to Aristarchus?', *Journal for the History of Astronomy*, 1985, Vol.16, No. 1, pp.37–42, accessed at adsabs.harvard.edu/full/1985JHA....16...37G, 16 May 2018. The quote is from Archimedes (*c*.216 BCE), *The Sand Reckoner*.

Helmasperger Notarial Instrument, 1455, accessed at www.gutenberg-digital.de/gudi/eframes/index.htm, 6 April 2018.

Hickok, J., 'Eight problems for the mirror neuron theory of action understanding in monkeys and humans', *Journal of Cognitive Neuroscience*, 2009, Vol.21, No.7, pp.1229–43, accessed at www.mitpressjournals.org/doi/full/10.1162/jocn.2009.21189?url_ver=Z39.88-2003&rfr_id=ori%3Arid%3Acrossref.org&rfr_dat=cr_pub%3Dpubmed.

Hinchliffe, J., 'Handwriting a struggle for more children, expert says', www.abc.net.au/news/2016-08-03/touch-screens-impacting-on-kids-writing-skills-therapist-says/7683054.

Hirsch, *Printing, Selling and Reading 1450–1550*, 1967, accessed at www.historyofinformation.com/expanded.php?id=3691.

Hodder, I., 'Catalhoyuk: the leopard changes its spots. A summary of recent work', *Anatolian Studies*, 2014, 64 pp.1–22.

Hodgson, D., Watson, B., 'The visual brain and the early depictions of animals in Europe and South-East Asia', *World Archaeology*, 2015, Vol. 47, No. 5, pp.776–91.

Holden, B.A., et al., 'Global prevalence of myopia and high myopia and temporal trends from 2000 through 2050', *Ophthalmology*, 2016, Vol.123, No.5, pp.1036–42, accessed at www.aaojournal.org/article/S0161-6420%2816%2900025-7/fulltext, 5 October 2018.

Holy Congregation for the Index, *Decree of the Holy Congregation for the Index Against Copernicanism*, 1616, accessed at inters.org/decree-against-copernicanism-1616, 12 June 2018.

Howe, N., *Forbes*, 2015, www.forbes.com/sites/neilhowe/2015/07/15/why-millennials-are-texting-more-and-talking-less/#3cde0b4e5576.

Hull, J., *Notes on Blindness* (2016), accessed at www.arte.tv/en/videos/069250-000-A/radio-hull and royalparks.org/history, 16 July 2016.

International Society of Aesthetic Plastic Surgeons, *Global Statistics 2017*, www.isaps.org/wp-content/uploads/2018/10/2017-Global-Survey-Press-Release-Demand-for-Cosmetic-Surgery-Procedures-Around-The-World-Continues-To-Skyrocket_2_RW.pdf.

Jakubiak, B.K., 'Affectionate touch to promote relational, psychological and physical wellbeing in adulthood: a theoretical model and review of the research', *Personality and Social Psychology Review*, 2016, Vol.21, No.3, pp.228–52, accessed at journals.sagepub.com/doi/10.1177/1088868316650307.

Keifer, M., Schuler, S, Mayer, C., Trumpp, N.M., Hille, K., Sachse, S., 'Handwriting or typewriting? The influence of pen- or keyboard-based writing training on reading and writing performance in preschool children', *Adv Cogn Psychol*, 2015, Vol.11, No.4, pp.136–46.

Kohli, P. et al., 'The association between olfaction and depression: a systematic review', *Chem Senses*, 2016, Vol. 41, No. 6, pp.479–486. doi: 10.1093/chemse/bjw061, accessed at www.ncbi.nlm.nih.gov/pmc/articles/PMC4918728, 21 January 2019. www.washingtonpost.com/lifestyle/food/what-is-dude-food-anyway-we-asked-the-experts-and-they-fired-away/2015/06/19/91da5fdc-15b5-11e5-89f3-61410da94eb1_story.html.

Kronfeld-Schor, N., Dominoni, D., de la Iglesia, H., Levy, O., Herzog, E.D., Dayan, T., Helfrich-Forster, C., 'Chronobiology by moonlight', *Proc R Soc B*, 2013, Vol.280.

Lamb, T.D., 'Evolution of phototransduction, vertebrate photoreceptors and retina', *Progress in Retinal and Eye Research*, Vol. 36 (Amsterdam: Elsevier, 2013), pp.52–119.

Lamb, T.D., Collin, S.P., Pugh, E.N., 'Evolution of the vertebrate eye: opsins, photoreceptors, retina and eye cup', *Nature Reviews Neuroscience*, 2007, Vol. 8, pp.960–76.

Lapidus, I.M., 'Cities and societies: a comparative study of the emerenge of urban civilisation in Mesopotamia and Greece', *Journal of Urban History*, Vol.12, No.3, pp.257–92 (Thousand Oaks, CA: Sage Publications, 1986), accessed 10 March 2016.

Leeming, David, 'Flood' entry, *The Oxford Companion to World Mythology* (Oxford: Oxford University Press, 2004) p.138, blog.britishmuseum.org/who-was-ashurbanipal/?_ga=2.217768542.950196694.1540219698-2130113507.1539634091.

Lekich, J., 'Lillian Gish – first lady of the silent screen', *Globe and Mail* (Vancouver), 24 October 1986.

'Letter from Gregory I to Abbott Mellitus', *Epistola*, 1215–16, accessed at sourcebooks.fordham.edu/source/greg1-mellitus.txt, 17 October 2017.

Levaco, R., 'Censorship, ideology, and style in Soviet cinema', *Studies in Comparative Communism*, 1984, Vol.XVII, Nos.3 and 4, pp.173–83, accessed at JSTOR.org.

Lewis, J.G., 'Smells ring bells: how smell triggers memories and emotions: brain anatomy may explain why some smells conjure vivid memories and emotions', *Psychology Today*, 2015, accessed at www.psychologytoday.com/gb/blog/brain-babble/201501/smells-ring-bells-how-smell-triggers-memories-and-emotions.

Livingstone, M., 'What art can tell us about the brain', 2015, lecture recorded at University of Michigan, accessed at www.youtube.com/watch?v=338GgSbZUYU, 13 August 2016.

Marshall, D., *Recollections of Edison*, 1931, Christopher Howe, Boston, p.60, cited in McPartland, D.S. *Almost Edison*.

Marshall, M., 'Oxygen crash led to Cambrian Mass Extinction', *New Scientist*, 2011, www.newscientist.com/article/dn19916-oxygen-crash-led-to-cambrian-mass-extinction.

Matthews, W., 'An Historical Sketch of the Origin, Progress and Present State of Gas-Lighting', 1827, Rowland Hunter, London. Cited in *The London Magazine*, 1827, Hunt and Clark, London, Vol. 9, p.515.

Mayne, J. (ed.), *Charles Baudelaire, The Mirror of Art*, 1955, Phaidon Press Limited, London, accessed at www.csus.edu/indiv/o/obriene/art109/readings/11%20baudelaire%20photography.htm.

McCouat, P., Early Influences of Photography on Art, Part 2, *Journal of Art in Society*, 2012, accessed at www.artinsociety.com/pt-2-photography-as-a-working-aid.html.

McGann, J. et al., 'Poor human olfaction is a 19th-century myth', *Science*, 2017, Vol.356, No.6338, accessed at science.sciencemag.org/content/356/6338/eaam7263, 21 January 2019.

McPartland, D.S., *Almost Edison: How William Sawyer and Others Lost the Race to Electrification*, 2006, Doctoral thesis UMI, No.3231946; accessed via ProQuest on 25 July 2016.

Mercado, M. et al., 'Trends in emergency department visits for nonfatal self-inflicted injuries among youth aged 10 to 24 years in the United States, 2001–2015', *Journal of the American Medical Association*, 2017, Vol.318, No.19, pp.1931–33, accessed at jamanetwork.com/searchresults?q=mercado&allJournals=1&Se archSourceType=1&exPrm_qqq={!payloadDisMaxQParser%20 pf=Tags%20qf=Tags^0.0000001%20payloadFields=Tags%20 bf=}%22mercado%22&exPrm_hl.q=mercado, Center for Disease Control Analysis of Emergency Dept Admissions.

Miner, E.D., Neptune, *Encyclopedia Britannica*, 2006, accessed at www. britannica.com/place/Neptune-planet/Neptunes-discovery.

Mojtobaj. R. et al., 'National trends in the prevalence and treatment of depression in adolescents and young adults', *Pediatrics*, 2016, Vol.138, No.6, accessed at pediatrics.aappublications.org/ content/143/1?current-issue=y, 20 January 2019.

Mommsen, T.E., 'Petrarch's conception of the 'Dark Ages', *Speculum*, 1942, Vol.17, No.2, pp.226–42, accessed at www.jstor.org/ stable/2856364, 1 November 2018, p.227.

Mongillo, P., Bono, G., Regolin, L., Marinelli, L., 'Selective attention to humans in companion dogs, Canis familiaris', *Animal Behaviour*, 2010, Vol. 80, No. 6, pp.1057–63.

Morgan, C. et al., 'Incidence, clinical management, and mortality risk following self-harm among children and adolescents: cohort study in primary care', *British Medical Journal*, 2017, Vol.359, accessed at www. bmj.com/content/359/bmj.j435120/1/19.

Morris, E., 'Which came first, the chicken or the egg', *New York Times*, 2007, opinionator.blogs.nytimes.com/2007/09/25/which-came-first-the-chicken-or-the-egg-part-one.

Morriss-Kay, G.M., 'The evolution of human artistic creativity', *Journal of Anatomy*, Vol.216, No.2.

Morse, R., *The Collected Writings of Benjamin Franklin and Friends*, 2004, Wright Center for Innovation in Science Teaching, Tufts University, Medford. www.tufts/edu/as/wright_center/personal_pages/bob_m/ franklin_electricity_screen.pdf.

Mortimer, Ian, 'The mirror effect: how the rise of mirrors in the fifteenth century shaped our idea of the individual', *Lapham's Quarterly*, 2016, accessed at www.laphamsquarterly.org/roundtable/ mirror-effect and fandomania.com/tv-review-mythbusters-8-27-presidents-challenge.

Mujcic, R. et al., 'Evolution of well-being and happiness after increases in consumption of fruit and vegetables', *Am J Public Health*, 2016, Vol.106, No.8, pp.1504–10, accessed at www.ncbi.nlm.nih.gov/pmc/ articles/PMC4940663, 21 January 2019.

Museum of the History of Science, Oxford University, MHS Collection Database Search, accessed at www.mhs.ox.ac.uk/collections/

imu-search-page/record-details/?TitInventoryNo=62524&querytype
=field&thumbnails=on////////&irn=3704.

NASA map of cumulative lightning strikes, 1995–2013, eoimages.
gsfc.nasa.gov/images/imagerecords/85000/85600/lightning_lis_
otd_1995-2013_lrg.jpg.

NASA website, exoplanets.nasa.gov/the-search-for-life/life-signs www.
space.com/40831-future-mars-rovers-search-alien-life.html.

NASA website, www.nasa.gov/mission_pages/hubble/science/star-v1.
html.

National Candles Association website, candles.org/history, accessed 16
July 2016.

Newzoo research, accessed at venturebeat.com/2018/09/11/
newzoo-smartphone-users-will-top-3-billion-in-2018-hit-3-8-
billion-by-2021, 18 January 2019.

Nichols, J.G. (ed.), 'Chronicle of the Grey Friars of London, 1852',
Camden Society Old Series, Vol.53, pp.53–78. *British History Online*,
accessed at www.british-history.ac.uk/camden-record-soc/vol53/
pp53-78, 5 April 2018.

Nikitina, N., Bronner-Fraser, M., Sauka-Spengler T., 'Sea
Lamprey Petromyzon marinus: A model for evolutionary and
developmental biology', *Cold Spring Harbor Protocols* (2009).

Nilsson, D.-E., 'The functional basis of eye evolution', *Visual
Neuroscience*, Vol. 30 (2013), accessed at journals.cambridge.org/
action/displayFulltext?type=1&fid=8889789&jid=VNS&volumeId=
30&issueId=1-2&aid=8889787&bodyId=&membershipNumber=&s
ocietyETOCSession=, 15 September 2015.

Nilsson, D.-E., Pegler, S., 'A Pessimistic estimate of the time required
for an eye to evolve', *Biological Sciences*, 2004, Vol. 256, No.1345,
pp.53–58, accessed at dx.doi.org/10.1098/rspb.1994.0048.

Norman, J., *Jeremy Norman's History of Information*, accessed at www.
historyofinformation.com.

Number of TV Households in America, 1950–1978, accessed at www.
tvhistory.tv/facts-stats.htm, 5 January 2019. web.archive.org/
web/20070831035110/, www.freetv.com.au/Content_Common/
pg-History-of-TV.seo.

Obituaries, *The Athenaeum*, No. 3515, 9 March 1895, p.314.

Ofcom, *Communications Market Report*, 2018, accessed at www.
ofcom.org.uk/__data/assets/pdf_file/0022/117256/CMR-2018-
narrative-report.pdf.

Ofcom, *Digital Day 2016*, accessed at stakeholders.ofcom.org.uk/
binaries/research/cross-media/2016/Digital_Day_2016_Overview_

of_findings_charts.pdf. Includes email and instant messaging but not the huge volume of messaging that goes through social media platforms.

Office of National Statistics figures, accessed at www.ons.gov.uk/peoplepopulationandcommunity/birthsdeathsandmarriages/deaths/bulletins/suicidesintheunitedkingdom/2017registrations#suicide-patterns-by-age.

Ovid, *Metamorphoses Book III*, trans Kline, A.S. (Independently published, 2014), accessed at ovid.lib.virginia.edu/trans/Metamorph3.htm, 15 January 2016.

Parkes, M.B., Introduction to Peter Ganz (ed), *The Role of the Book in Medieval Culture*, Colection symposium papers (Belgium: Brepols, 1986).

Pausas, J.G., Keeley, J.E., 'A burning story: the role of fire in the history of life', *BioScience*, 2009, Vol. 59, No. 7, pp.593–601, www.biosciencemag.org.

Pepys, S., *Diary of Samuel Pepys*, 1667, accessed at www.pepysdiary.com/diary/1667/26/05, 1 June 2018. Also www.visioneng.com/resources/history-of-the-microscope.

Pew Research Centre, *Millennials, a Portrait of Generation Next, Confident. Connected. Open to Change*, 2010, www.pewsocialtrends.org/files/2010/10/millennials-confident-connected-open-to-change.pdf.

Pierre, J., 'The Narcissism epidemic and what we can do about it', 2016, *Psychology Today*, www.psychologytoday.com/gb/blog/psych-unseen/201607/the-narcissism-epidemic-and-what-we-can-do-about-it.

Plato, *Phaedrus*, trans. Jowett, B., via Project Gutenberg, accessed at www.gutenberg.org/files/1636/1636-h/1636-h.htm, 29 October 2018.

Pohl, A., McGuire, J., Toobie, A., 'P6 9 Laser Diode Another Day', *Journal of Physics Special Topics*, 2013, accessed at journals.le.ac.uk/ojs1/index.php/pst/article/download/2153/2057.

Pope Jean Paul, *Allocution of the Holy Father Jean Paul*, 1992, accessed at bertie.ccsu.edu/naturesci/cosmology/galileopope.html

Powell, B., 'Why was the Greek alphabet invented? The epigraphical evidence', *Classical Antiquity*, Vol.8, No.2 (Berkeley, CA: University of California Press, 1989), p.346, www.jstor/stable/25010912.

Power, S. et al., 'Sleepless in school? The social dimensions of young people's bedtime rest and routines', *Journal of Youth Studies*, 2017, accessed via Taylor & Francis.

Prince, S. Hensley, W., 'The Kuleshov effect: recreating the classic experiment', *Cinema Journal*, 1992, Vol.31, No.2 (Winter, 1992), pp.59–75, accessed at JSTOR.org, 2 January 2019.

Prins, H.E.L., Bishop, J., 'Edmund Carpenter: explorations in media and anthropology', *Visual Anthropology Review*, 2002, Vol. 17, No. 2, pp.110–31.

Putnam, R., 'Bowling alone: America's declining social capital', *Journal of Democracy* 6:1, January 1995, pp.65–78.

Rabin, S., 'Nicolaus Copernicus', *The Stanford Encyclopedia of Philosophy*, 2015 (autumn 3025 Edition), Edward N. Zalta (ed.), accessed at plato.standford.edu/archives/fall 2015/entries/Copernicus, 16 May 2018.

RGS Archives, Rawlinson, *Personal Adventures*, cited in Adkinds, 2004.

Richards, J., 'Cinemagoing in worktown: regional film audiences in 1930s Britain', *Historical Journal of Film, Radio and Television*, 1994, Vol.14, No.2, p.147.

Richardson, R., 'Why do wolves howl?', *slate.com*, 2014, accessed at www.slate.com/blogs/wild_things/2014/04/14/why_do_wolves_howl_wolves_do_not_howl_at_the_moon.html, 14 May 2018, also accessed at www.nationalgeographic.org/media/wolves-fact-and-fiction, 14 May 2018.

Riefenstahl, L., 1935, cited in Sontag, S., 1975, *Fascinating Fascism*, accessed at docs.google.com/viewer?a=v&pid=sites&srcid=Z-GVmYXVsdGRvbWFpbnxmYXNjaXNtYW5kbWFzY3VsaW5pdHl8Z3g6NDAxMjhmNmEzN2M1OTM3M3, 3 January 2019.

Riva, M.A., Arpa, C., Gioco, M., 'Dante and asthenopia: a modern visual problem described during the Middle Ages', *Eye*, 2014, Vol.28, p.498.

Robertson, D., Davidoff, J., Davies, I.R.L., Shapiro, R.L., 'Colour categories and colour acquisition in Himba and English', in Pitcham, N., Biggam, C.P., *Progress in Colour Studies: Vol. II Psychological Aspects* (Amsterdam: John Benjamins Publishing, 2006).

Robinson, A., 'Thomas Young and the Rosetta Stone', *Endeavour*, 2007, Vol.31, No.2, accessed at sciencedirect.com, 12 March 2016.

Roebroeks, W., Villa, P., 'On the earliest evidence for habitual use of fire in Europe', 2011. *Proceedings of the National Academy of Sciences of the United States of America*, Vol.108, No.13, pp.5209–14.

Ross, C.F., Kirk, E.C., 'Evolution of eye size and shape in primates', *Journal of Human Evolution*, 2007, Vol. 52, pp. 294–313.

Rossell, D., *The Magic Lantern*, Published in von Dewitz et al. (ed.), 2002, *Ich Sehe was, was du nicht siehst! Sehmaschinen und Bilderwelten*,

Steidl/MuseumLudwig/Agfa Fotohistoram, Cologne, accessed at www.academia.edu/345943/The_Magic_Lantern.

Rothman, J., 'When Bigotry Paraded through the streets', *The Atlantic*, 2016, December 2016, accessed at www.theatlantic.com/politics/archive/2016/12/second-klan/509468, 29 December 2018.

Rouse, M.A., Rouse, R.H., 'Backgrounds to print: aspects of the manuscript book in northern Europe of the fifteenth century', *Authentic Witnesses: Approaches to Medieval Texts and Manuscripts* (Notre Dame, IN: University of Notre Dame Press, 1991).

Rung, A.M., Joseph Saxton, Inventor and Pioneer Photographer, *Pennsylvania History Journal*, 1940, Vol.7, No.3 (Pennsylvania University Press, July 1940), pp.153–58, , accessed at www.jstor.org/stable/27766416.

Sandgathe, D.M., Dibble, H.L., Goldberg, P., McPherron, S.P., Turq, A., Niven, L., & Hodgkins, J., Timing of the appearance of habitual fire use, *Proceedings of the National Academy of Sciences of the United States of America*, 2011, Vol.108, No.29, E298.

Schmandt-Besserat, D., 'Writing systems', 2008, in Pearsall, D. (ed.), *Encyclopedia of Archaeology* (Oxford: Elsevier Science & Technology, 2008), accessed at 0-search.credoreference.com.wam.city.ac.uk/content/entry/estarch/writing_systems/0, 15 March 2016.

Science Daily, 'One in five young people lose sleep over social media', *ScienceDaily*, 16 January 2017, accessed at www.sciencedaily.com/releases/2017/01/170116091419.htm.

Selzer, L.J., et al., 'Instant messages vs. speech: hormones and why we still need to hear each other', *Evolution of Human Behavior*, 2012, Vol.33, No.1, pp.42–45, accessed at www.ncbi.nlm.nih.gov/pmc/articles/PMC3277914, 21 January 2019.

Shakespeare, W., *A Midsummer Night's Dream*, Act III, Scene ii, *c.*1595.

Shakespeare, W., *Hamlet*, Act II, Scene ii, Line 535: 'Follow him friends, we'll hear a play tomorrow'.

Sherwin, J.C. et al., 'The association between time spent outdoors and myopia in children and adolescents: a systematic review and meta-analysis', *Opthalmology*, 2011, Vol.119, No.10, pp.2141–51.

Shimelmitz, R., Kuhn, S.L., Jelinek, A.J., Ronen, A., Clark, A.E., Weinstein-Evron, M., 'Fire at will: the emergence of habitual fire use 350,000 years ago', *Journal of Human Evolution*, 2014, Vol.77, pp.196–203.

Siddique, H., 'Mental health disorders on rise among children', 2018, *The Guardian*, 22 November 2018, accessed at www.theguardian.

com/society/2018/nov/22/mental-health-disorders-on-rise-among-children-nhs-figures.

Sines, G., Sakellarakis, Y.A., 'Lenses in antiquity', *American Journal of Archaeology*, 1987, Vol.91, No.2, pp.191–96, accessed at www.jstor.org/stable/505216, 7 November 2018.

Singham, M., 'The Copernican myths', *Physics Today*, 2007, Vol.60, No.12, p.48.

Smith, A.M., 'Alhacen on refraction: a critical edition, with English translation and commentary, of book 7 of Alhacen's "De Aspectibus", the Medieval Latin version of Ibn al-Haytham's "Kitāb al-Manāzir", Vol.2', *Transactions of the American Philosophical Society*, 2010, Vol.100, No.3, Section 2.

Smith, T.J. et al., 'A window on reality: perceiving edited moving images', *Current Directions in Psychological Science*, 2012, Vol. 21, No. 2, pp.107–113, accessed journals.sagepub.com/doi/abs/10.1177/0963721412437407?journalCode=cdpa.

Snowdon, C., *Closing Time: Who's Killing the British pub?* 2016, report for the Institute of Economic Affairs, accessed at www.iea.org.uk/sites/default/files/publications/files/Briefing_Closing%20time_web.pdf.

Spear, B., 'Let there be light! Sir Joseph Swan and the incandescent light bulb', 2013, *World Patent Information*, Vol. 35, No. 1, Elsevier, pp.38–41.

Sperling, J., 'Multiples and reproductions: prints and photographs in nineteenth century England – visual communities, cultures and class', in Kromm J. (ed.), *A History of Visual Culture* (Oxford: Berg Publishers, 2010).

Stevenson, R.L., 'A plea for Gas Lamps', *Virginibus Puerisque and Other Papers*, 1893, Charles Scribner's Sons, New York.

Stowe, H.B., *Pogaunc People: Their Lives and Loves* (New York, NY: Howard and Hulbert, 1878), p.230.

Suger, 1140, translation of original manuscript, accessed at www.learn.columbia.edu/ma/htm/ms/ma_ms_gloss_abbot_sugar.htm and www.smithsonianmag.com/smart-news/the-first-nativity-scene-was-created-in-1223-161485505/.

Sugovic, M., Turk, P., Witt, J.K., 'Perceived distance and obesity: it's what you weigh, not what you think', *Acta Psychol (Amst)*, 2016, Vol.165, pp.1–8.

Swerdlow, N.M., 'Copernicus's derivation of the heliocentric theory from Regiomontanus's eccentric models of the second inequality of the superior and inferior planets', *Journal for the History of Astronomy*,

2017,Vol.48, No.1, pp.33–61, accessed at journals.sagepub.com/doi/abs/10.1177/0021828617691203, 17 May 2018.

Taylor Parker, S., Mitchell, R.W., Boccia M.L., 'Self Awareness in Animals and Humans', *Developmental Perspectives* (Cambridge: Cambridge University Press, 1994).

Taylor Redd, N., 'Who discovered Uranus, and how do you pronounce it?', *space.com*, 2018, accessed at www.space.com/18704-who-discovered-uranus.html.

Tomasello, M., Hare, B., Lehmann, H., Call, J., 'Reliance on head versus eyes in the gaze following of great apes and human infants: the cooperative eye hypothesis', *Journal of Human Evolution*, 2007, Vol. 52, pp. 314–320.

Triumph of the Will film viewed at archive.org/details/TriumphOfTheWillHQ

Turner, H.A. Jnr, *Germany from Partition to Reunification*, 1992, Yale University Press.

Twenge, J.M., 'iGen: why today's super-connected kids are growing up less rebellious, more tolerant, less Happy –and completely unprepared for adulthood – and what that means for the rest of us', *The Atlantic*, 2017, www.theatlantic.com/magazine/archive/2017/09/has-the-smartphone-destroyed-a-generation/534198.

Unknown, 'Fine arts: the Daguerre secret', *The Literary Gazette; and Journal of the Belles Lettres, Arts, Sciences, &c, 1839*, No.1179 (Saturday, 24 August 1839): pp.538–39.

Unknown Editor, New Discovery, 1839, *Blackwood's Edinburgh Magazine* (Edinburgh and London) 45: 281 (March 1839): pp.382–91. Accessed via Gary W. Ewer (ed.), *The Daguerreotype: an Archive of Source Texts, Graphics, and Ephemera*, www.daguerreotypearchive.org.

Van Helden, A., 'The invention of the telescope', *Transactions of the American Philosophical Society*, 1977, Vol.67, No.4, pp.1–67.

Various, *Galileo Trial: 1616 Documents*, accessed at douglasallchin.net/galileo/library/1616docs.htm.

Various, 'Optics, instruments and painting, 1420–1720: reflections on the Hockney–Falco thesis', *Early Science and Medicine*, 2005, Vol.10, No.2.

Victoria and Albert Museum, *Sea and Sky: Photographs by Gustave Le Gray 1856–1857*, exhibition notes, 2003, accessed at www.vam.ac.uk/content/articles/s/gustave-le-grey-exhibition, 3 December 2018. www.telegraph.co.uk/news/2017/11/16/leonardo-da-vincis-salvator-mundi-sells-450-million-342-million, accessed 2 December 2018.

Walcott, C.D., *Field Diary Notes*, 1909, accessed via Royal Ontario Museum website, burgess-shale.rom.on.ca/en/history/discoveries/02-walcott.php.

Weissner, P.W., 'Embers of society: firelight talk among the Ju/'hoansi Bushmen', *Proceedings of the National Academy of Sciences of the United States of America*, 2014, Vol.111, No.39, pp.14027–35, www.pnas.org/cgi/doi/10.1073/pnas.1404212111.

Whipple, F.H., *Municipal Lighting, Detroit*, 1888, cited in Schivelbusch (1995), p.127.

Whipps, H., 'How smallpox changed the world', *Live Science*, 2008, accessed at www.livescience.com/7509-smallpox-changed-world.html, 15 October 2017.

White, C., 'The smell of commerce', 2011, *The Independent*, 16 August 2011, www.independent.co.uk/news/media/advertising/the-smell-of-commerce-how-companies-use-scents-to-sell-their-products-2338142.html, www.addmaster.co.uk/scentmaster/scentmaster-technology.

Wood, D.W., *The Arrival of the Daguerreotype in New York*, monograph for The American Photographic Historical Society (New York), 1995, accessed at www.midley.co.uk/daguerreotype/newyork.htm, 25 July 2018.

Woolley, S., promotional material for his film *Their Finest* about the Film Unit in the Second World War, 2017, based on Lissa Evans' novel *Their Finest Hour and a Half* (Cambridge: Black Swan, 2010), accessed at www.pressreader.com/uk/daily express/20170413/2817 32679350947.

www.businesswire.com/news/home/20180910005394/en/Global-Male-Grooming-Products-Market-2018-2023.

www.cosmopolitan.co.uk/beauty-hair/news/a25314/uk-eyebrow-trends-statistics-2014.

www.groominglounge.com/blog/marketing-stuff/men-vs-women-who-spends-more-time-grooming.

www.history.com/news/ask-history/what-is-the-origin-of-the-handshake.

www.latin-dictionary.net/search/latin/mirare, 1 February 2016.

Xiong, S. et al., 'Time spent in outdoor activities in relation to myopia prevention and control: a meta-analysis and systematic review', *Acta Ophthalmol*, 2017, Vol.95, No.6, pp.551–66, accessed at www.ncbi.nlm.nih.gov/pmc/articles/PMC5599950, 15 January 2019.

Xu, Y., Zhu, S.-W., Li, Q.-W., 'Lamprey: a model for vertebrate evolutionary research', *Zoological Research*, 2016, Vol. 37, No. 5, pp.263–69.

Young, T., 'An account of some recent discoveries', *Hieroglyphical Literature and Egyptian Antiquities* (London: John Murray, 1823) pp.xiv–xv.

Zax, D., 'Galileo's revolutionary vision helped usher in modern astronomy', *Smithsonian Magazine*, 2009, accessed at www.smithsonianmag.com/science-nature/Galileos-Revolutionary-Vision-Helped-Usher-In-Modern-Astronomy-34545274/.

Zhao, F., Bottjer, D.J., Hu, S., Yin, Z., Zhu, M., 'Complexity and diversity of eyes in Early Cambrian ecosystems', *2013 Scientific Reports*, Vol. 3, No. 2751.

Zimecki, M., 'The lunar cycle: effects on human and animal behaviour and psychology', *Postepy Hig Med Dosw*, 2006, accessed at www.phmd.pl/api/files/view/1953.pdf.

INDEX